U.S. ARMY
HAND-TO-HAND
COMBAT

U.S. ARMY HAND-TO-HAND COMBAT

DEPARTMENT OF THE ARMY

SKYHORSE PUBLISHING

Skyhorse Publishing books may be purchased in bulk at special discounts for sales promotion, corporate gifts, fund-raising, or educational purposes. Special editions can also be created to specifications. For details, contact the Special Sales Department, Skyhorse Publishing, 307 West 36th Street, 11th Floor, New York, NY 10018 or info@skyhorsepublishing.com.

Visit our website at www.skyhorsepublishing.com.

10

Library of Congress Cataloging-in-Publication Data

U.S. Army hand-to-hand combat / Department of the Army.
     p. cm.
  ISBN 978-1-60239-782-8
  1. Hand-to-hand fighting--United States--Handbooks, manuals, etc. 2. United States. Army--Physical training--Handbooks, manuals, etc. 3. Soldiers--Training of--United States--Handbooks, manuals, etc. I. United States. Dept. of the Army. II. Title: US Army hand-to-hand combat. III. Title: United States Army hand-to-hand combat.
  U167.5.H3U17 2009
  355.5'480973--dc22

                                200902456

Printed in China

Field Manual⎫　　DEPARTMENT OF THE ARMY
No. 21–150　⎭　Washington 25, D. C., *June 14, 1954*

HAND-TO-HAND COMBAT

*This manual supersedes FM 21–150, 30 June 1942.

CHAPTER 1

INTRODUCTION

1. Purpose and Scope

This manual is written to teach you hand-to-hand combat. It describes the various blows, holds, footwork, armwork, and other maneuvers used to disable or kill an enemy in hand-to-hand fighting. It also explains how to use all available objects as weapons. Hand-to-hand combat stresses simple, aggressive tactics. You can subdue an opponent only through offensive measures.

2. Necessity for Training

The average soldier, if trained only in the use of his basic weapon, loses his effectiveness if his weapon fails to fire or if he should lose or break it. With a knowledge of hand-to-hand combat and the confidence and aggressiveness to fight hand-to-hand, the soldier is able to attack and dispose of his opponent. Training in hand-to-hand combat is also useful for night patrols and other occasions when silence is required. This type fighting is taught to soldiers in rear areas as well as those in front lines because of the threat of infiltration, airborne attacks, and guerilla warfare.

CHAPTER 2

FUNDAMENTALS OF HAND-TO-HAND COMBAT

3. General

Five fundamentals are used as a guide in learning hand-to-hand combat. These fundamentals are making full use of any available weapon; attacking aggressively by using your maximum strength against your enemy's weakest point; maintaining your balance and destroying your opponent's; using your opponent's momentum to advantage; and learning each phase of all the movements precisely and accurately before attaining speed through constant practice.

4. Using Available Weapons

a. When fighting hand-to-hand, your life is always at stake. The use of any object as a weapon, therefore, is necessary to help subdue your enemy. You can make your opponent duck or turn aside by throwing sand or dirt in his face or by striking at him with an entrenching tool, a steel helmet, or a web belt. When no object is available, just the pretense of throwing something may cause an enemy to flinch and cover up. When he does this, you must take advantage of his distraction to attack aggressively with but one purpose in mind—TO **KILL**.

b. If no objects are available to use as a weapon, you must make full use of your natural weapons. These are—

    (1) *The knife edge of your hand.* Extend your fingers rigidly so the little finger edge of your hand is as hard as possible (fig. 1). Keep your thumb alongside your forefinger.

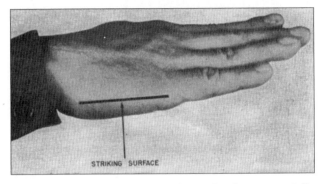

STRIKING SURFACE

Figure 1. With the knife edge of your hand, you can strike many killing and disabling blows.

    (2) *The fingers folded at the second knuckles.* The average fist covers an area of about eight square inches. The fingers folded at the second knuckles gives a striking surface of about two square inches, producing a sharper, more penetrating blow. Keep your thumb tightly against the forefinger to stiffen your hand and keep your wrist straight (fig. 2).

    (3) *The protruding second knuckle of your middle finger.* Fold the middle finger at the second knuckle and wedge the second knuckles of your two adjacent fingers into

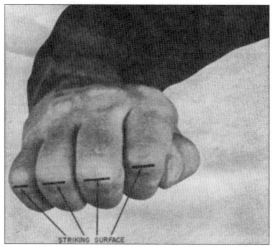

STRIKING SURFACE

Figure 2. Blows delivered with the fingers folded at the second knuckles produce sharp penetration.

its sides. Keep the end of the thumb over the fingernail of your middle finger and keep your wrist straight (fig. 3).

(4) *The heel of your hand.* Fold your fingers at the second knuckles and force the back of your hand toward the wrist to make the heel of your hand as solid as possible (fig. 4). You can deliver a more damaging blow with the heel of your hand than with your fist.

(5) *The little finger edge of your fist.* Form a fist. When using the little finger edge of your fist as a weapon, strike blows in the same motion as when using an ice pick (fig. 5).

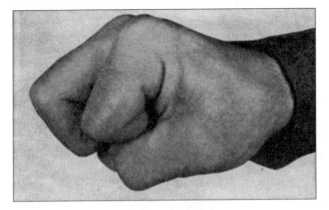

Figure 3. Dangerous blows can be delivered to vulnerable points with the protruding second knuckle of the middle finger.

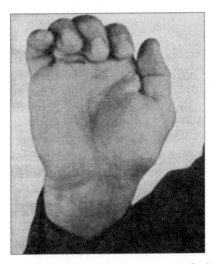

Figure 4. The heel of the hand is particularly effective when attacking parts of the face.

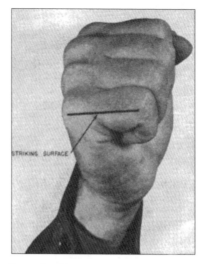

Figure 5. Powerful blows with the little finger edge of the fist can easily kill an opponent.

(6) *Your boot.* For most kicks, use the outside or inside edge of your boot rather than the toe. This provides a much larger striking surface with which to attack small, exposed bony areas (fig. 6).

(7) In addition to the natural weapons already mentioned, you can use your elbows, knees, head, shoulders, and teeth to disable an opponent.

5. Maximum Strength Against Weakest Point

Using maximum strength against your enemy's weakest point is an axiom of war that equally applies to combat between two individuals. In every situation, some extremely vulnerable area of your opponent is open for attack. By aggressively as-

Figure 6. The inside or outside edge of the boot is more effective than the toe, which may slip off small areas.

saulting these vulnerable areas, using the maximum strength offered by your position, you can gain a quick victory. Attacking rather than defending is the keynote because only through the use of offensive tactics are you able to dispose of your enemy.

6. Balance

a. Keeping your own balance, while causing your opponent to lose his, is an important essential of successful fighting. Assume the guard position when engaging your opponent (fig. 7). This position is similar to a boxer's crouch and enables you to react rapidly and move in any direction. Spread your feet about shoulder's width apart, with your left heel generally on line with your right toe. If you are left handed, reverse this position and bring your left foot behind your right foot. Bend your body forward at the waist and at the knees slightly. Hold the hands at face level and slightly in front of it. Extend and join your fingers, with the thumbs along the forefingers and the palms facing inward. Face your opponent squarely. The guard position offers you the best balanced position you can obtain before

Figure 7. The guard position offers good balance and good all-around protection.

closing with your opponent. You will improve your sense of balance and learn to destroy your opponent's balance after closing with him by practicing the maneuvers presented in this manual.

b. When fighting, keep your feet spread laterally to maintain balance. Destroy your opponent's mental balance by growling and yelling as you strike at him.

7. Momentum

Using your opponent's momentum to your own advantage is another fundamental. Always assume that your opponent is stronger than you and never oppose him directly in a test of strength. Instead, utilize his momentum and strength to overcome him. Examples of using your opponent's momentum are tripping him, side stepping as he rushes you, or ducking his blow.

8. Accuracy and Speed

You will have little time to stop and think when engaging in hand-to-hand combat. Therefore, your actions must be automatic. At the beginning, learn each phase of each movement separately and accurately, putting the stress on precision alone. As you progress, work for speed through constant practice. Speed is essential to the successful employment of most of the maneuvers outlined in this manual.

CHAPTER 3

VULNERABLE POINTS

Section I. INTRODUCTION

9. General

a. Vulnerable points are areas of the body that are particularly susceptible to blows or pressure. Knowledge of these points and how to attack them, plus aggressiveness and confidence, will enable you to attack and quickly disable or kill the enemy you meet in hand-to-hand combat.

b. When you are attacking an opponent, your first reaction is probably to strike him on the jaw with your closed fist. This is one of the poorest ways to fight. A better attack is to strike your opponent across the bridge of his nose with the knife edge of your hand. This type blow could easily break the thin bone in his nose, causing extreme pain and temporary blindness. A severe blow could drive bone splinters into his brain and cause instant death. These actions must be performed *without hesitation* and *with aggressiveness.*

10. Body Regions

The body is divided into three regions: The head and neck, the trunk, and the limbs. Here is a list of the major vulnerable points of each region—

Head and neck	*Trunk*	*Limbs*
a. Eyes.	*a.* Groin	*a.* Instep
b. Nose	*b.* Solar plexus	*b.* Ankle
c. Adam's apple	*c.* Spine	*c.* Knee
d. Temple	*d.* Kidney	*d.* Shoulder
e. Side of neck	*e.* Collar bone	*e.* Elbow
f. Nape	*f.* Floating ribs	*f.* Wrist
g. Upper lip	*g.* Stomach	*g.* Fingers
h. Ears	*h.* Armpit	
i. Base of throat		
j. Chin		

11. Caution

Only a small amount of pressure or a light blow is needed to injure or kill a man when attacking some of the vulnerable points. It is important, therefore, to strike very light blows in training when learning how to attack these points. When thoroughly trained, you may add a little more force to your blows; but still remember the vulnerability of the area being attacked in order not to injure your training partner.

Section II. HEAD AND NECK

12. Eyes

There are various ways to blind an opponent. One is to drive your index and middle fingers, formed into a **V**, into your opponent's eyes (fig. 8). Keep your fingers stiff and your wrist firm. You can also use the second knuckles of two adjacent fingers in a sharp thrust at the eyes. The eyes can be gouged out by using your thumbs or fingers.

13. Nose

When attacking the nose, strike a forceful blow with the knife edge of your hand across the bridge (fig 9). This blow can easily break the thin bone,

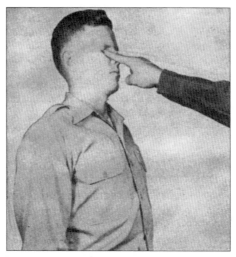

Figure 8. *The fingertips driven forcefully into the eyes can easily blind an opponent.*

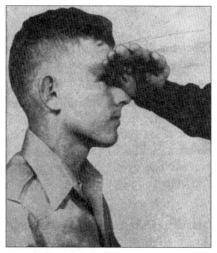

Figure 9. *A forceful blow to the bridge of the nose will knock an opponent out of action.*

causing your opponent extreme pain and temporary blindness. A very sharp blow could drive bone splinters into your opponent's brain and kill him instantly. You can also use the outside edge of your closed fist. When fighting at close quarters, attack the nose by hitting the bottom of it an upward blow with the heel of your hand.

14. Adam's Apple

Attack the Adam's apple with the knife edge of your hand (fig. 10). A severe blow can result in death by severing the windpipe. A lesser blow is painful and causes your opponent to gag. The Adam's apple is also vulnerable to attack with the fist, toe, or knee, depending upon your opponent's position. Squeezing it or pulling it outward with the fingers and thumb is another method you can use.

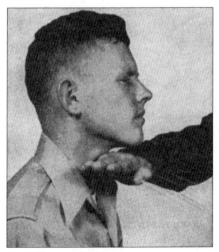

Figure 10. A severe blow to the Adam's apple with the knife edge of the hand can kill.

15. Temple

A blow to the temple can easily kill or cause a concussion. The bone structure at this spot is weak and an artery and a large nerve lie close to the skin. Attack the temple with the knife edge of your hand or with the outside edge of your closed fist (fig. 11). A jab with the point of your elbow can also be used. If you succeed in knocking your opponent down, kick his temple with your toe.

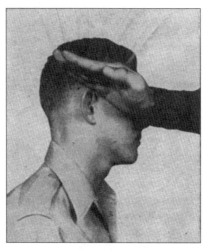

Figure 11. The temple is a weak part of the body. A forceful blow here will usually kill an enemy.

16. Nape

A blow with the knife edge of your hand to your opponent's nape ("rabbit punch") could easily kill him by breaking his neck (fig. 12). The outside edge of your fist can also be used. Use this blow if your opponent charges low and his hands are not guarding the upper regions of his body. If you succeed in

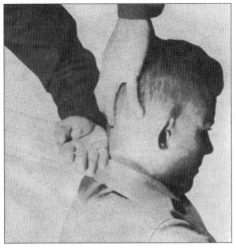

Figure 12. The "rabbit punch" (blow to nape).

knocking your opponent down, kick his nape with your toe, stomp it with your heel, or strike it with the knife edge of your hand.

17. Side of Neck

One way to knock your opponent unconscious is to deliver a sharp blow with the knife edge of your hand to the side of the neck, below and slightly to the front of the ear (fig. 13). You can deliver it in two ways: A backhand delivery with the palm down or a forward slash with the palm up. This type blow causes unconsciousness by shocking the jugular vein, the carotid artery, and the vagus nerve. It is not particularly dangerous.

18. Upper Lip

A vulnerable part of the face is the upper lip, just below the nose, where the nose cartilage joins the

298982°—54——2

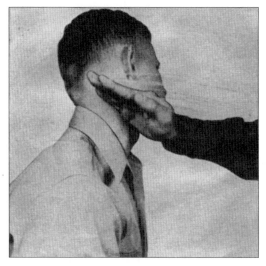

Figure 13. A blow to the side of the neck will not kill an enemy, but if delivered forcefully, it can cause unconsciousness.

bone. The nerves here are close to the skin. This area can be attacked by delivering a sharp blow with the knife edge of your hand at a slightly upward angle (fig. 14). A very sharp blow can cause unconsciousness. A lesser blow causes extreme pain. A jab with the second knuckles of your fingers can also be used.

19. Ears

Cup your hands and clap them simultaneously over your opponent's ears (fig. 15). This is a dangerous blow and may burst his ear drums, cause nerve shock, or result in possible internal bleeding. A sharp enough blow can cause a brain concussion and death.

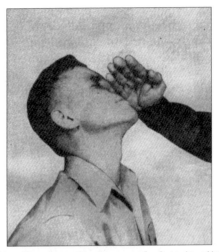

Figure 14. The upper lip is a good place to attack when fighting close-in.

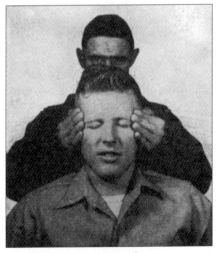

Figure 15. Clapping cupped hands over an enemy's ears can kill.

20. Base of Throat

One way to break an opponent's hold on you is to quickly thrust one or two extended fingers into the small indentation at the base of his throat (fig. 16). The blow is painful and causes him to gag and cough. Severe injury could result if the thin layer of skin at this point is pierced.

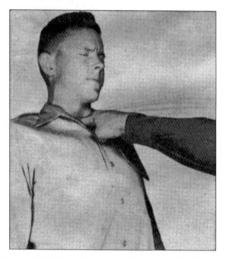

Figure 16. Jabbing a finger or fingers into the base of an opponent's throat causes him to loosen a hold.

21. Chin

An effective blow can be delivered to your opponent's chin with the heel of your hand, which is better than a closed fist (fig. 17). You may break a bone in your hand by using your fist.

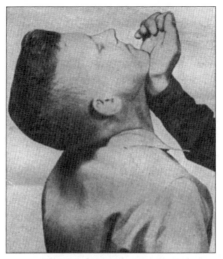

Figure 17. Striking an opponent on the chin with the heel of the hand is better than striking him with a fist.

Section III. TRUNK

22. Groin

When closing with an opponent, keep in mind that one of the best points to attack is the groin.   Attack it by kicking up forcefully with your knee (knee-lift) (fig. 18).   You can also use your closed fist, the knife edge of your hand, grasping fingers, a toe kick, or a heel stomp.

23. Solar Plexus

The solar plexus is at the bottom of the rib cage, just beneath the breast bone. To attack this area, thrust sharply with the second knuckle of the protruding middle finger (fig. 19).   This method permits sharp penetration and is, therefore, more effective than striking this small target with the fist

*Figure 18.   Attacking the groin is one of the most effective
methods of subduing an opponent.*

*Figure 19.   A blow to the solar plexus with the protruding
knuckle of the middle finger permits sharp penetration.*

or the knife edge of your hand. Any sharp blow to the solar plexus causes extreme pain and may either bend your opponent forward or drop him to his knees. Death may result from a severe blow.

24. Spine

The spinal column houses the spinal cord and a blow here can cause derangement of the column, resulting in paralysis or death. If you succeed in knocking your opponent down, a blow with your knee, your elbow, the heel of your shoe, or a toe kick can easily kill or seriously injure him (fig. 20). The best place to strike this blow is three or four inches above the belt line where the spine is least protected.

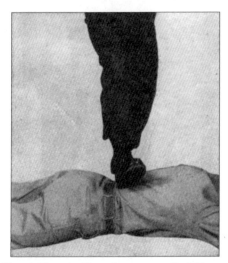

Figure 20. A blow to the spine can easily kill or seriously injure an opponent.

25. Kidney

Certain large nerves, branching from the spine, are very close to the skin surface over the kidneys. A blow here can rupture the kidney and cause severe nervous shock or death unless the victim receives immediate medical attention. To attack this area, use the knife edge of your hand (fig. 21). Other effective blows can be delivered with fingers folded at the second joints, the outside edge of your fist, the knee, or a toe kick.

Figure 21. Attacking the kidney.

26. Collar Bone

A forceful blow delivered straight down on the collar bone at the side of the neck with the knife edge of your hand can fracture the bone and cause your opponent to drop to his knees (fig. 22). Another way of attacking this point, and a particularly good way if your opponent is shorter than you, is to drive your elbow down into the collar bone.

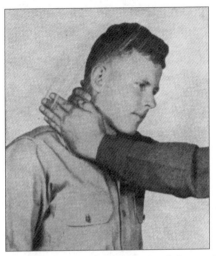

Figure 22. The knife edge of the hand is an effective weapon to use against the collar bone.

27. Floating Ribs

Attack the floating ribs from either the front or rear, but, if possible, strike the blow to your opponent's right side. The liver is located here just below the ribs, and the blow causes terriffic shock to this organ. Attack this area with the knife edge of your hand (fig. 23), the outside edge of your fist, the knuckles folded at the second joints, the heel, the toe, or the knee.

28. Stomach

A blow to your opponent's stomach with your knuckles folded at the second joints causes him to loosen his hold on you (fig. 24). If he bends for-

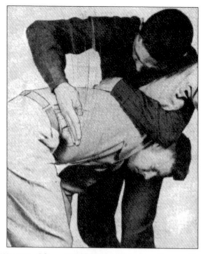

Figure 23. Attacking the floating ribs.

Figure 24. To break a hold, hit an opponent in his stomach with the knuckles folded at the second joints.

ward, strike him in the face with your knee or deliver a "rabbit punch" to his nape. The knuckle blow gives sharper penetration than a blow with the elbow or fist. A toe kick or a kneelift can also be used and could cause serious injury.

29. Armpit

A large nerve is close to the skin in the armpits. A blow to this area causes severe pain and temporary partial paralysis. If you succeed in knocking your opponent down, attack the armpit with a toe kick (fig. 25).

Figure 25. If you knock your opponent down, kick his armpit to cause temporary partial paralysis.

Section IV. LIMBS

30. Instep

The small bones of the instep can easily be broken with a stomp, causing severe pain to your opponent as well as limiting his movement. When facing your opponent, deliver a stomp with the edge of your left boot to his left instep or with the edge of your right boot to his right instep (fig. 26). This type delivery protects your groin area as you turn. Follow the blow to the instep with a blow to the ankle. Kick your opponent sharply on the outside of his ankle with the outside edge of your boot. Do not use a toe kick because it may slip off your opponent's ankle without doing damage.

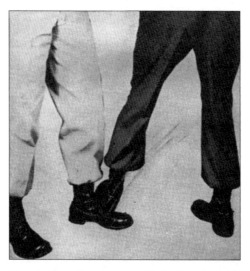

Figure 26. A stomp to an opponent's instep can easily break the bones here.

31. Knee

Kick your opponent's knee or kneecap with the edge of your boot (fig. 27). The blow will tear ligaments and cartilage, causing him extreme pain and affecting his mobility. If you succeed in getting

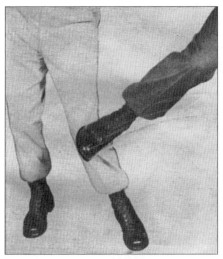

Figure 27. A kick to an opponent's knee will hinder his mobility.

behind your opponent, a sharp toe kick to the back of his knee will penetrate his flesh and injure the nerves.

32. Shoulder

After you knock your opponent down, you can easily dislocate his shoulder by twisting his arm behind his back and dropping on his shoulder with your weight on one knee (kneedrop) (fig. 28). When you are in position to do this, you also can fall on your opponent's spine, causing paralysis or immediate death.

33. Elbow

The elbow joint is a comparatively weak part of the body and a forceful blow can dislocate it. Grasp your opponent's wrist or forearm and pull it behind him, stiffening his whole arm (fig. 29). As

Figure 28. A kneedrop to an opponent's shoulder will dislocate this part of the body and make his arm useless.

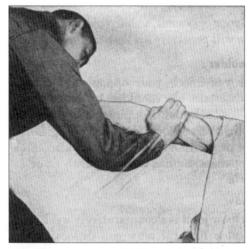

Figure 29. Once an opponent's elbow is broken, he ceases to be dangerous.

you do this, give his elbow a sharp blow with the heel of your hand. The knife edge of your hand or your knee can also be used.

34. Wrist

Bending the wrist excessively in any direction causes extreme pain. Use a wristlock when attacking this area. Place both your thumbs on the back of your opponent's hand. Bend the wrist at a right angle to his forearm (fig. 30). You can control your opponent when you get him in this position.

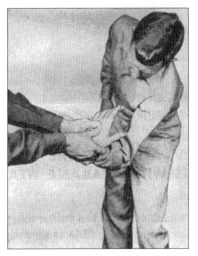

Figure 30.   A wristlock produces severe pain.   An opponent can be controlled in this position.

35. Fingers

To break an underarm hold around your waist from the rear, grasp any one of your opponent's fingers with one hand while securing his wrist with the other (fig. 31). Push down on his wrist and, at

the same time, bend his finger back toward his wrist,
This will break his finger.

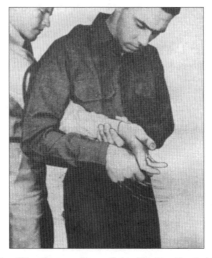

Figure 31. The fingers are vulnerable to attack if an oppo-
nent has secured a waist hold.

Section V. ATTACKING THE VULNERABLE POINTS WITH AVAILABLE WEAPONS

36. General

You can attack many of the vulnerable points more effectively by using many objects as weapons.

37. Bayonet Hilt and Tent Peg Knob

Grasp the bayonet or the tent peg so the hilt of the bayonet or the knob of the tent peg protrudes from the little finger edge of your hand (fig. 32).

38. Homemade Blackjack

You can make a blackjack by placing wet sand or a bar of soap in a sock. Tie a knot in the sock just

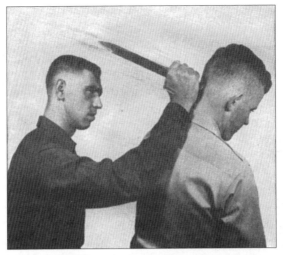

Figure 32. The hilt of the bayonet can be used to silence an enemy sentry.

above the sand or the soap. When attacking an opponent, strike him on the back of his head (fig. 33).

39. Blunt Objects

By striking your opponent between the shoulder blades on his spine with a blunt object, you can knock him out noiselessly. You can use the blunt end of a hand axe or the butt of a rifle (fig. 34). A blow with the toe of the rifle or the edge of the axe will kill your opponent instead of stunning him.

40. Tent Rope

You can strangle an unsuspecting enemy sentry by using a tent rope or a piece of wire (pars. 50 and 51).

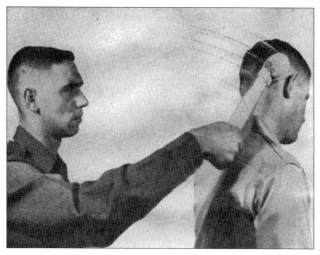

Figure 33. The homemade blackjack is used on many night patrols.

Figure 34. To stun an enemy, use the butt end of the rifle. To kill him, use the toe.

41. Other Weapons

When you find yourself unarmed, you can, on most occasions, find a piece of equipment, a rock, a stick, or a club to use as a weapon. By using these and by attacking viciously, concentrating on your opponent's vulnerable points, you can quickly kill or disable him.

CHAPTER 4

KNIFE ATTACK

42. General

A knife, properly employed, is a most deadly weapon. You can use it on patrols when silence is necessary to prevent an outcry by an enemy sentry, or you can use it for close-in fighting when your rifle or carbine is not available.

43. Grip

To grip the knife properly, lay it diagonally across the outstretched palm of your hand. Grasp the small part of the handle next to the cross guard with your thumb and forefinger. Your middle finger encircles the knife over the handle at its largest diameter (fig. 35). With the knife held in this manner, it is easily maneuvered in all directions. You can control the direction of the blade by a combination movement of the forefinger and middle finger and a turning of the wrist. When the palm is turned up and you are holding the knife in your right hand, you can slash to the right or left. When the palm is turned down, you can also slash in either direction. You can thrust when the palm is held either up or down. When the knife makes contact, it is grasped tightly by all fingers.

44. Stance

When engaging in a knife attack, you are in a crouch with your left hand forward and the knife

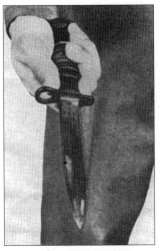

Figure 35. A good grip on a knife is essential for control.

held close to your body at the right hip (fig. 36).
Your outstretched left hand acts as a guard, a foil,
or a parry, and it helps create the opening for a slash
or a thrust. You may also use your left hand to distract your opponent's attention by waving it in his
face, by throwing something, or by making sudden
darting motions toward him. When you are in this
type crouch, your flexed knees provide extreme mobility and you have good balance. In the crouch,
you are also able to protect your midsection and
throat area.

45. Where To Attack

a. When a man is attacked from the front with a
knife, he instinctively tries to protect his stomach
and throat. If he is wounded in one of these places,
his fear is so great that he may forget to defend him-

Figure 36. A proper stance provides mobility and good protection.

self further. His opponent, therefore, can easily kill him.

b. You can attack the throat with either a thrust or a slash. The thrust is the most effective if the knife is driven into the base of the throat just below the Adam's apple (fig. 37). This type blow cuts the jugular vein and results in instant death. A slash to either side of the neck cuts the carotid artery, which carries blood to the brain. Your opponent will die from loss of blood within a few seconds.

c. A thrust (fig. 38) combined with a slash to the stomach produces great shock. Your enemy will be stunned and will forget to defend himself. You can then deliver a killing blow. A deep wound in the stomach causes death if the wound is unattended.

d. A thrust to the heart (fig. 39) causes instant death. This spot, however, is difficult to hit because

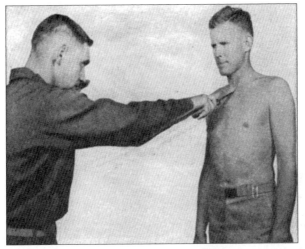

Figure 37. An enemy will die immediately if a knife is thrust into the base of his throat.

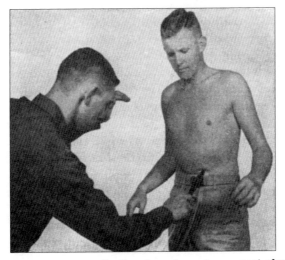

Figure 38. A thrust to the stomach produces great shock.

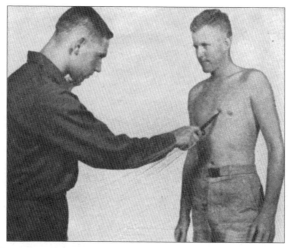

Figure 39. A thrust to the heart causes instant death. The heart, however, is protected by ribs.

of the protecting ribs. Usually, a hard thrust will slip off the rib and penetrate the heart.

e. A slash to the wrist (fig. 40) will sever the radial artery, causing death within two minutes. This type attack is excellent if your opponent attempts to grasp your clothing or arm. The radial artery is only one-quarter inch below the surface of the skin. Unconsciousness results in about 30 seconds.

f. A slash to the upper arm just above the inside of the elbow (fig. 41) cuts the brachial artery and causes death within 2 minutes. This artery is about one-half inch below the skin surface. Unconsciousness occurs in about 15 seconds.

g. A slash to the inside of the leg near the groin (fig. 42) severs the arteries there and makes that limb useless.

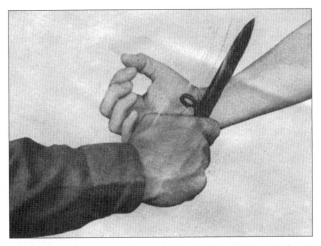

Figure 40. A slash to the wrist cuts the radial artery and will kill an enemy within two minutes.

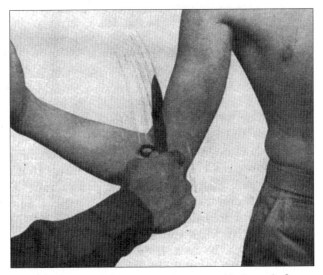

Figure 41. The upper arm is vulnerable to a slash.

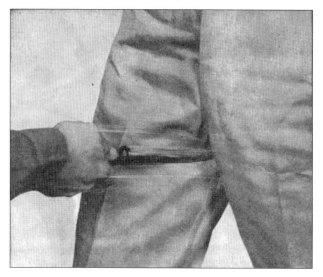

Figure 42. A slash to the inside of the leg will immobilize an opponent.

46. Attack From the Rear

a. To attack an enemy from the rear, you launch your attack immediately on reaching a position not less than 5 feet from him. Thrust the knife into his right or left kidney and simultaneously grasp his mouth and nose with the other hand (fig. 43). After a short interval, withdraw the blade, slashing as you do so, and cut his throat. The thrust to the kidney produces great shock and causes internal hemorrhage and death.

b. A thrust into the side of the neck (fig. 44 ①) is also effective when you want to maintain silence. A slash across the neck from the rear (fig. 44 ②) severs the windpipe and jugular vein.

Figure 43. A thrust to the kidney from the rear disposes of an enemy silently.

c. The subclavian artery is approximately 2½ inches below the surface between the collar bone and shoulder blade (fig. 45). Attack this spot with a thrust by gripping the knife as you would an ice pick. As you withdraw the knife, slash to make the wound as large as possible. This artery is difficult to hit, but once it is cut, the bleeding cannot be stopped and your opponent will lose consciousness within seconds. Death will follow rapidly.

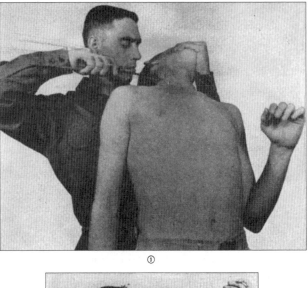

①

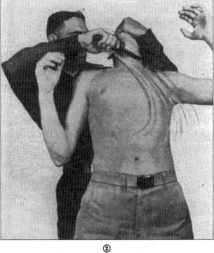

②

Figure 44. A thrust or a slash to the neck from the rear kills an enemy quickly and silently.

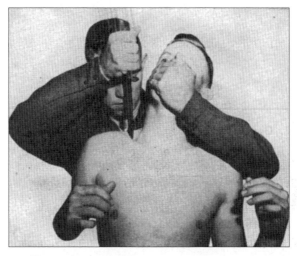

Figure 45. Attacking the subclavian artery.

CHAPTER 5

SILENCING SENTRIES

47. General

Knowing the methods of silencing sentries will enable you to maintain silence and surprise and kill an unsuspecting man from his rear quickly, quietly, and efficiently.

48. Helmet Neck Break

Grasp the front rim of your opponent's helmet with your right hand. At the same time, place your left forearm against the back of his neck and place your left hand on his right shoulder (fig. 46 ①). Holding firmly to the front rim of your opponent's helmet, pull his helmet up, back, and down and press your left forearm forward (fig. 46 ②). Your left forearm, under the back edge of his helmet, acts as a fulcrum against which his neck is broken. This method is possible only when your opponent's helmet strap is fastened underneath his chin.

49. Helmet Smash

If you see that your opponent's helmet strap is not fastened or should you discover this when attempting the helmet neck break, silence the man with a helmet smash. Pull your opponent's helmet quickly from his head. While doing this, grasp his collar with your other hand, jerking him off balance to his rear (fig. 47 ①). Then smash the helmet to the back of his head or at the back of his neck (fig. 47 ②).

①

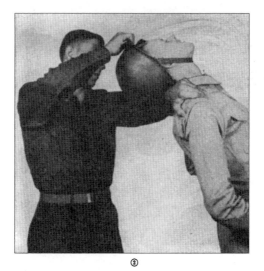

②

Figure 46. The helmet neck break.

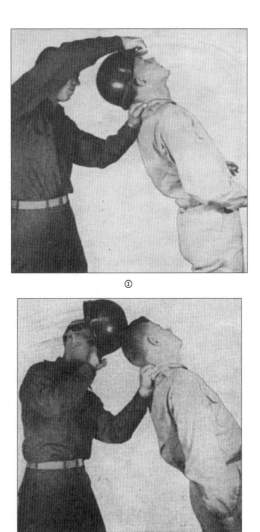

①

②

Figure 47. Use the helmet smash when an opponent's helmet strap is not fastened.

Your opponent may have a chance to yell when this method is used.

50. Strangulation With Wire or Cord

For this method of strangling an opponent to death, you need a piece of flexible wire or a piece of cord about three feet in length. Approach the enemy from his rear, holding an end of the wire or cord in each hand. Toss the wire or cord over his neck from his left and place the heel of your left hand, still holding the end of the wire or cord, on his shoulder near his nape. At the same time, place your knee in the small of the man's back and pull back on the cord or wire forcefully with your right hand while pushing with the left hand (fig. 48). If this is done quickly, your opponent cannot cry out. You can tie the ends of the rope or wire around two short sticks for a better hold.

Figure 48. A piece of wire or cord is a silent weapon.
298982°—54——4

51. Two Hand Loop

Hold an end of the wire or cord in each hand. Place your left forearm across the back of your opponent's neck as shown in figure 49 ①. Swing your right arm over your opponent's head from his right, looping the wire or cord in front of his throat. Complete the loop and jerk your arms sharply in opposite directions, tightening the loop and strangling your opponent (fig. 49 ②). Quick application of this method prevents your opponent from crying out. You can cause unconsciousness or death, depending on the force used and the length of time the hold is applied.

52. Other Methods

Other methods of silencing sentries can be found throughout this manual under other chapter headings. To help you find these methods, a list of them and the chapters and paragraphs where they are found are given below.

    a. Knife attack, chapter 4.

        (1) Thrust to kidney (par. 46*a*).

        (2) Thrust to side of neck (par. 46*b*).

        (3) Throat slash (par. 46*b*).

    b. Available weapons, chapter 3, section V.

        (1) Striking an opponent on his spine with a blunt object to stun him and with a sharp object to kill (par. 39).

        (2) Use of the homemade blackjack (par. 38).

    c. Natural weapons, chapter 3, section II. Striking an opponent on the base of the skull with the knife edge of your hand or the little finger edge of your fist (par. 16).

    d. Holds, chapter 7, section I.

        (1) Taking a man down from his rear (par. 69).

        (2) Locked rear strangle hold (par. 73).

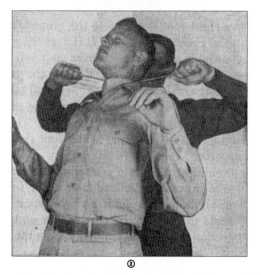

Figure 49. The two hand loop method of strangulation is quick and silent.

CHAPTER 6

FALL POSITIONS AND THROWS

Section I. SIDE FALL POSITIONS

53. General

You must learn various fall positions before you attempt the throws that are taught in unarmed combat. Constant practice in these positions will enable you to be thrown without being injured.

54. Left Side Fall Position

Figure 50 shows the left side fall position. The check points for this position are—

a. Your right foot is driven to the ground, taking up the initial shock of the fall. It strikes the ground before your body and is crossed over your left leg at the knee. The sole of your foot is flat on the ground.

b. Your left arm is the "beating" arm and takes up additional shock. It is extended along the ground, palm down, at an angle of 45° to the body. This arm makes contact with the ground at the same time your "shock absorber" foot does.

c. Your chin is tucked into your chest. Keep your neck tense to prevent your head from being injured.

d. Your right arm is folded across your chest. This prevents injury to your right elbow and offers some protection to the head and body from the blows of your opponent.

Figure 50. Every soldier must learn how to fall properly to prevent injury.

e. The entire left side of your body makes solid contact with the ground. To relax your left leg, bend it slightly to prevent it from being injured.

f. The check points for the right side fall position are the same as those for the left side fall position. Simply substitute the words "right" for "left" and "left" for "right."

55. Practicing the Falls

a. Figure 51 illustrates a method of practicing the fall positions. Your training partner assumes a position on his hands and knees. Lie with your back across his back and position your legs and right arm into the proper position for the fall. Fold your left arm across your chest.

Figure 51. A basic way to practice the left side fall position is to fall off a training partner's back in the proper position.

b. Watch the ground over your left shoulder and swing your left arm forcefully to your left, rolling off your partner's back. Your left hand and right foot strike the ground first, taking up the initial shock of the fall. The right side fall position can be practiced in a similar manner.

56. Advanced Falling Practice

a. Start from the guard position. Take several steps forward to build up momentum. When your left foot strikes the ground, kick your right leg vigorously into the air. At the same time, thrust with your left foot so that your body is propelled into the air, feet first, and is parallel to the ground.

While in midair, twist your body 90° to the left and, at the same time, cross your right foot over your left leg at the knee. The sole of your right foot faces toward the ground so it will strike the ground first, thus taking up the initial shock of the fall. While still in the air, extend your left arm in front and at a 45° angle to your body. Your palm is down to take up the additional shock of the fall and to prevent injury to the elbow. Tuck your chin into your chest and fold your right arm across your chest. This prevents injury to your head and right arm when you make contact with the ground. When you strike the ground, you have good contact along the entire left side of the body, your right foot having absorbed most of the shock. The right side fall position can be practiced in a similar manner.

b. Start from the right side fall position on the ground. Push off the ground vigorously with your left foot and right arm in order to turn your body to the left side. During the turn, pull your knees against your chest and thrust them vigorously upward and outward at an angle of 15° so that your entire body is lifted off the ground. Once you are in the air, twist 90° to your left, assuming the left side fall position. By completely clearing the ground when changing from the right to the left side fall position, you obtain practice in absorbing landing shock. The right side fall position can be practiced from the left side in a similar manner.

Section II. OVER SHOULDER THROW FALL POSITION

57. From a Standing Position

Use this fall position when your opponent throws you over his shoulder. To practice this fall position, start from the guard position. Take several steps forward to build up momentum, and then assume a squatting position as if going into a forward roll. Place your hands between your knees, with the palms flat on the ground. Arch your back and tuck your chin into your chest to keep your head from striking the ground (fig. 52 ①). Roll forward in a somersault. At the peak of the forward roll, drive the soles of the feet to the ground about shoulder's width apart, keeping your lower legs at a 90° angle to the ground. This takes up the initial shock of the fall. Keep your stomach muscles tightened so your

①

Figure 52. The over shoulder throw fall position.

②

Figure 52—Continued.

buttocks will not strike the ground when you land. At the same time your feet strike the ground, slap both hands to the ground. The arms are fully extended and the palms down, forming a 45° angle to your body. This slapping motion gives you contact with the ground along both arms and across the shoulders, taking up the fall's additional shock (fig. 52 ②). After completing the fall, check the following points:

a. Are the soles of both feet flat on the ground?

b. Are the lower legs at a 90° angle to the ground?

c. Are the buttocks well off the ground?

d. Is the chin tucked into the chest?

e. Are the shoulders and arms flat on the ground, palms down, with the arms at a 45° angle to the body?

58. From the Ground

To practice the over shoulder throw fall position without coming to a standing position, do the following:

a. Lie down on the ground, draw your knees up to your chest, fold your arms across your chest, and rock into a sitting position.

b. Roll backward as if going into a backward roll.

c. At the peak of your backward roll (when your shoulders touch the ground), thrust your feet vigorously upward and outward at an angle of 15°, raising your body completely off the ground.

d. While in midair, tighten your stomach muscles and strike the soles of your feet to the ground.

e. Make contact with the ground with your shoulders, arms, and soles of the feet all at the same time.

Section III. BASIC THROWS

59. General

At times in hand-to-hand combat, you have to throw your opponent to the ground before you can attack a vulnerable part of his body. Three basic throws used are the right hip throw, the over shoulder throw, and the reverse hip throw. Variations of these throws can be used and new ones taught after you have learned the basic ones. An additional basic throw, the leg hock, is not described in this manual.

a. Speed is the primary factor in throwing an opponent in combat. In training, however, strive for precision and accuracy. Do each phase of the throws with deliberate action. Once you have thoroughly learned the throws, work for speed through constant practice.

b. In the beginning, your partner should offer no resistance. He should cooperate and permit you to execute the throw while he concentrates on assuming a good fall position.

c. The three throws described in this section may be executed from either side simply by substituting the words "right" for "left" and "left" for "right."

60. Right Hip Throw

a. Start the right hip throw from the guard position, facing your opponent. Place your left foot in front of and slightly inside of your opponent's left foot. At the same time, strike your opponent vigorously on his right shoulder with the heel of your left hand and grasp his clothing here (fig. 53 ①). This blow knocks him off balance.

b. Pivot to your left 180° on the ball of your left foot. During your pivot, place your right arm around your opponent's waist and jerk him forward forcefully with both arms, driving his midsection into your buttocks. This maneuver bends your opponent over your right hip at his waist and leaves him partially suspended in this position. At the completion of this maneuver, your buttocks are into your opponent's midsection, your right foot is in front of and slightly outside of his right foot, and your knees are bent (fig. 53 ②).

① Knock an opponent off balance by striking him forcefully on the right shoulder

② Suspend the opponent in midair by jerking forcefully forward with both arms

Figure 53. Right hip throw.

c. Straighten your legs quickly, thrusting your buttocks forcefully into your opponent's midsection. At the same time, bend forward at the waist and pull forward and down with both arms, driving your opponent to the ground (fig. 53 ③). Use your hip as a fulcrum, throwing the man over your right hip and not over the outside of your leg. At the completion of the throw, your opponent lands in the left side fall position. You are poised to deliver a blow to a vulnerable part of his body.

③ Be ready to deliver a killing blow to a vulnerable point

Figure 53. Right hip throw—Continued.

61. Over Shoulder Throw

a. Start the over shoulder throw from the guard position, facing your opponent. The first phase of this throw is identical to the first phase of the right hip throw and the foot work is identical throughout to that of the right hip throw. Place your left foot in front of and slightly inside of your opponent's left foot. At the same time, strike him vigorously on his right shoulder with the heel of your left hand and grasp his clothing here (fig. 53 ①).

b. Pivot to your left 180° on the ball of your left foot. Keep your right arm in position to protect your head and neck region until you near the completion of the pivot. Then reach up and grasp your opponent's clothing at his right shoulder with your right hand, gripping him here with a double hand hold. As you complete your pivot, pull your opponent forward and drive his midsection into your buttocks (fig. 54 ①). Your buttocks are directly in front of your opponent's hips, your right foot is in front of and slightly outside of your opponent's right foot, your elbows are as close to your body as possible, and your knees are bent.

c. Straighten your legs, bend at the waist, and pull downward with both hands (fig. 54 ②). This action will catapult him over your shoulder. Your partner assumes the over shoulder throw fall position as he strikes the ground.

62. Reverse Hip Throw

a. Start the reverse hip throw from the guard position, facing your opponent. Stand slightly closer to him than in the two previous throws. Take a long step forward with your left foot and place it slightly outside of and a few inches beyond your opponent's right foot. Most of your weight is supported on your left foot. At the same time, strike your opponent forcefully on his right upper arm with your left hand and grasp his arm at this point with that hand (fig. 55 ①). This blow causes him to lose his balance to the rear.

①

②

Figure 54. The over shoulder throw.

① Secure a good grip on the opponent's right arm

② Gain buttock to buttock contact and suspend the opponent in midair

Figure 55. Reverse hip throw.

b. Step around your opponent with your right foot and place it directly behind him. At the same time, encircle his waist with your right arm. Push your hips to your right as far as possible to gain buttock to buttock contact, and pull your opponen position on your right hip. At this time, lock your opponent's right arm into your side with your left elbow (fig. 55 ②).

c. Straighten your legs, use your right hip as a fulcrum, and slam the man to the ground (fig. 55 ③). Be sure to use your hip as a fulcrum and throw your

③ If the fall does not knock the man out, deliver a blow to a vulnerable point

Figure 55. Reverse hip throw—Continued.

298982°—54——5

opponent over your hip, not over the side of your right leg. Notice in figure 55 ③ that you retain the armlock on your opponent's right arm. Also notice that the man who was thrown has assumed the left side fall position.

Section IV. VARIATIONS

63. Variations of the Hip Throw

a. Figure 56 ① illustrates a variation of the left hip throw. Both of your opponent's arms are securely pinioned, his right arm with a single elbowlock and his left arm clasped at the elbow.

b. Figure 56 ② illustrates another variation. This time you grasp your opponent's right arm with both hands and again use your hip as a fulcrum.

c. In a third variation, place your right arm around your opponent's neck as you pivot, rather than around his waist. Your left hand locks your opponent's right arm (fig 56③).

64. Variations of the Over Shoulder Throw

a. Figure 57① illustrates a variation of the over shoulder throw. While facing your opponent, grasp his right wrist with your left hand. Then pivot to your left 180°, pulling him forward as you do so. Grasp his right upper arm with your right hand and throw him over your shoulder as described in paragraph 61.

b. Figure 57② illustrates another variation of the over shoulder throw. From a position facing your opponent, grasp his left lapel with your right hand. Maintain this hold and pivot 180° to the left, placing your right forearm under his right armpit as you complete the pivot. Grasp his right arm at the elbow as you execute your pivot.

① Pin both the opponent's arms

② Use the hip as a fulcrum

Figure 56. Variations of the hip throw.

③ Grasp the opponent around his neck

Figure 56. Variations of the hip throw—Continued.

① Use the opponent's arm for leverage

Figure 57. Variations of the over shoulder throw.

② Grasp the opponent by his lapel

③ Grasp the opponent by his hair

Figure 57. Variations of the over shoulder throw—
Continued.

c. Figure 57 ③ illustrates another variation. When your opponent attacks you from the rear, grasp his hair or lock both arms around his head and throw him over your shoulder.

65. Variations of the Reverse Hip Throw

a. Figure 58① illustrates a variation of the reverse hip throw. Instead of placing your right arm around your opponent's waist, get a strangle hold around his throat.

b. In this variation, grasp the hand of the arm which you placed around your opponent's throat as described in *a* above. This gives you a strangle hold (fig. 58②).

①

Figure 58. Variation of the reverse hip throw.

⊙

Figure 58. Variation of the reverse hip throw—Continued.

CHAPTER 7

HOLDS AND ESCAPES

Section I. HOLDS

66. General

The two purposes of a hold are—

a. To kill your opponent immediately by applying enough pressure to certain parts of the body.

b. To hold your enemy until you can follow through with a blow to a vulnerable part of the body.

67. Front Strangle Hold

The front strangle hold is particularly good against a low frontal attack. As your opponent charges, slap your left hand against his right shoulder to slow his momentum and slip your right forearm under his throat. Clamp his head under your arm. Clasp your left wrist with your right hand. Apply pressure by leaning backward and lifting with your right forearm (fig. 59 ①). You can choke your opponent to death in this position. Another way to execute this hold and one which acts more swiftly is illustrated in figure 59 ②. Grasp the knife edge of your right hand with the fingers of your left, pull forcefully toward your chest and, at the same time, lean backward. Properly executed, any strangle hold can cause unconsciousness in approximately seven seconds. Continued pressure will kill a man in less than 1 minute. *When applying this hold, keep the*

①

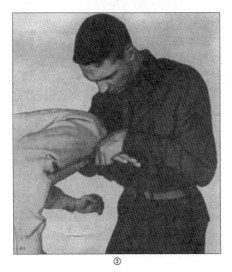

②

Figure 59. The front strangle hold is particularly good against a low frontal attack.

bony, inside edge of your forearm across your oppo-nent's Adam's apple for maximum effectiveness.

68. Side Collar Strangle Hold

Grip your opponent well back on his collar with both hands, palms down. Use the back of his collar for leverage and roll the second knuckles of your forefingers into the carotid arteries at the sides of his neck. Place both your thumbs below his Adam's apple, applying continuous pressure inward and upward (fig. 60). This hold is best used when your opponent is on the ground and unable to attack your groin. It causes unconsciousness and eventual death by stopping the flow of blood to the brain.

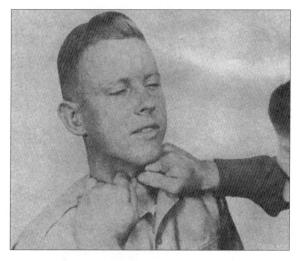

Figure 60. The side collar strangle hold stops the flow of blood to the brain and causes eventual death.

69. Taking a Man Down From His Rear

Figure 61 ① illustrates the start of the proper method of taking a man down from his rear. Your hands and foot make contact with your opponent's body simultaneously. Clap your hands down on his shoulders and, at the same time, pull backward. Kick the sole of the right foot forcefully against the back of your victim's knee joint (fig. 61 ②). This attack drops your victim to the ground instantly and places him in a position to be disabled or killed quickly. To knock your enemy unconscious, drive your knee to the base of his skull as he goes down.

70. Cross Collar Strangle Hold

To be effective, this hold must be executed on an individual who is wearing an open collar or who has open lapels on his coat or jacket. Cross your hands at the wrist and grasp the collar opening with your fingers on the inside and your thumbs on the outside. Pull strongly with your fingers and scissor your arm against your opponent's throat (fig. 62). He will drop to the ground unconscious. This strangle hold can also be executed from the rear. Cross your arms in front of your opponent's throat, seize his clothing at the neck, and press your arms into his throat by pulling tight.

71. Full Nelson

Execute this hold on your enemy from the rear. Place both your arms well up into your opponent's armpits and place your hands on the back of his head. Interlock your fingers (fig. 63 ①). Apply downward pressure on his head and upward pressure under his arms (fig. 63 ②).

①

②

Figure 61.   If an opponent has his back turned, he can easily be thrown to the ground and killed quickly.

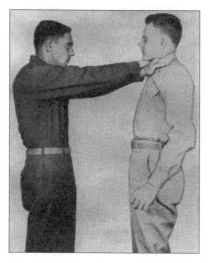

*Figure 62.   The cross collar strangle hold, executed properly,
causes unconsciousness.*

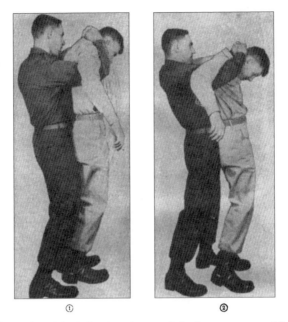

Figure 63.   The full nelson is executed when an opponent has his back turned.

72. Hammerlock

To execute this hold from the rear, grasp your opponent's hand or wrist and pull backward. Then, force his forearm up toward his head (fig. 64). By keeping your right hand at his right elbow and by continuing to force up on his arm, you can easily dislocate his shoulder. To execute a hammerlock when facing your opponent, grasp his right wrist with both hands. Pivot to your left 180°. During your pivot, raise your opponent's arm above your head and step beneath it and behind him at the completion of your turn.

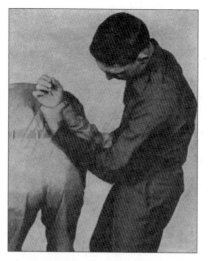

Figure 64. An opponent's shoulder can easily be dislocated by using the hammerlock.

73. Locked Rear Strangle Hold

Use the locked rear strangle hold when approaching your opponent from his rear. Place your left hand on the back of his head and, at the same time, cross your right forearm under his neck from the right (fig. 65 ①). Bring your right forearm to the left and lock it to your left upper arm (fig. 65 ②). In this position, push with your hand on the back of his head and lean forward. Enough pressure can break his neck. *Keep the inside, bony edge of the right forearm over your opponent's Adam's apple for best effect.*

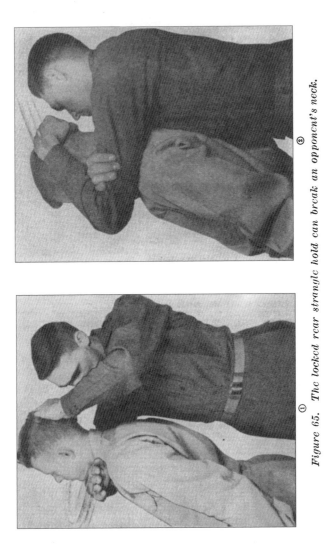

① ② *Figure 65. The locked rear strangle hold can break an opponent's neck.*

74. Double Wristlock

To execute a double wristlock, grasp your opponent's right wrist with your left hand (fig. 66 ①). Pass your right hand and arm over his right upper arm, under his bent elbow, and clasp your left wrist, completing the double wristlock (fig. 66 ②). Continue the movement by jerking his arm up and back into a twisting hammerlock (fig. 66 ③). Figure 66 ④ illustrates a variation of the double wristlock.

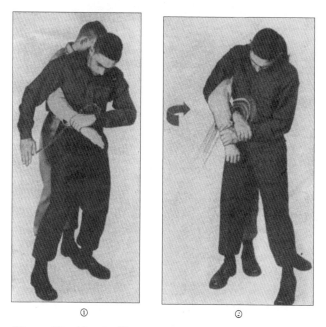

①   ②

Figure 66. The double wristlock may be continued into a hammerlock.

298982°—54——6

③   ④

Figure 66.   The double wristlock may be continued into a hammerlock—Continued

Section II. ESCAPE FROM HOLDS

75. General

If your opponent succeeds in getting a hold on you, you must break this hold before or immediately after he completes it. Bite, kick, or strike him at vulnerable points to help loosen or break the hold before he can apply pressure. By escaping from your opponent's grasp immediately, you can take the offensive again and attack him.

76. Escape From Choke Hold

When your opponent attempts to choke you (fig. 67 ①), use the arm swing to break his hold. Swing your arm over his arms in a forceful roundhouse blow. As you do this, pivot in the direction of your swing to get as much of your body weight behind your arm as possible (fig. 67 ②). This causes your opponent to loosen his hold. Be prepared to strike him across the face or the side of the neck with a backhanded blow with the knife edge of your hand before he recovers. This escape can also be used against a choke hold from the rear. Swing your arm and pivot around, facing your attacker as you swing.

77. Second Escape From Choke Hold

As your opponent gains the hold, clasp your hands together (fig. 68 ①). Grip the knife edge of your left hand with the fingers of your right, and tightly wrap the left thumb around the right thumb. Do not interlock your fingers. Drive your hands up between your opponent's arms, forcing him to loosen his hold (fig. 68 ②). From this position, smash your clasped hands on the bridge of his nose (fig. 68 ③), or grasp the back of his head and pull it down, meet-

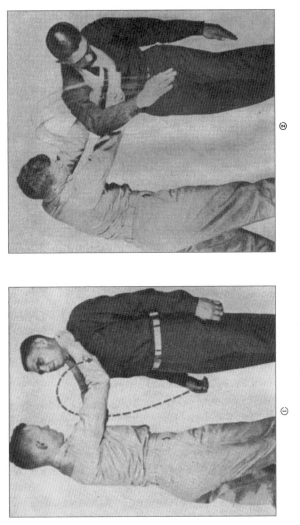

① ②

Figure 67. The choke hold is broken by a roundhouse arm swing.

①

②   ③

Figure 68. The choke hold can be broken and an opponent injured in one continuous motion.

ing it with a knee-lift. You can also separate your hands after breaking the hold and strike his collar bone with the knife edges of your hands.

78. Escape From Two-Hand Strangle Hold When Pinned Against Wall

Ordinarily, an opponent attempting to strangle you while your back is to a wall extends his arms, squeezes with his fingers, and pushes you against the wall (fig. 69 ①). To escape from this hold, place the heel of your right hand on his left elbow and the heel of your left hand on his right elbow. Apply pressure inward and away from you (fig. 69 ②). This prevents your opponent from using the power of his fingers and he cannot choke you. To drive him back, drive your knee or toe into his groin.

79. Escape From One-Arm Strangle Hold From the Rear

When your opponent attacks as shown in figure 70 ①, reach up with your left hand and grasp his clothing at his right elbow. Pull down on his elbow and, at the same time, tuck your chin into its crook so he cannot choke you. Grasp your opponent's right shoulder with your right hand (fig. 70 ②). Push backward with your buttocks against his midsection, retaining your hold on his upper arm and shoulder with both hands. Bend from the waist swiftly and throw your opponent over your head and to the ground (fig. 70 ③).

①   ②

Figure 69. Escape from two-hand front strangle hold when pinned against wall.

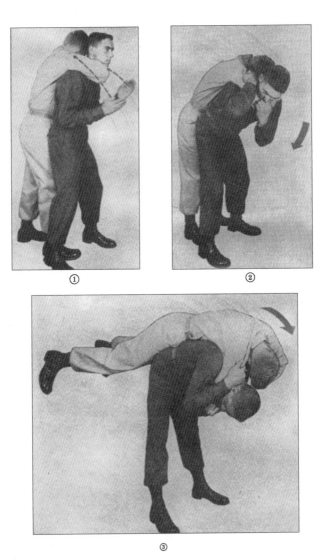

① ② ③

Figure 70. Escape from one-arm strangle hold from the
rear.

80. Escape From Front Overarm Bear Hug

When your opponent has grasped you around the body and pinned both your arms as shown in figure 71 ①, bring your thumbs into his groin, forcing his hips backward and leaving a space between your hips and his (fig. 71 ②). Pivot on your left foot and place your right foot outside of your opponent's right foot. Slip your right arm under his left armpit and grasp him across the back. Your left hand grasps his right upper arm, pulling it forcefully. Force your buttocks into his midsection and, at the same time, twist to your left. Lift with your right arm and pull with your left hand, throwing your opponent over your hip and to the ground (fig. 71 ③).

81. Escape From Overarm Rear Body Hold

When your opponent attacks as shown in figure 72 ①, loosen his grip by stepping on his instep or kicking his shins. Raise your elbows shoulder high and, at the same time, lower your body quickly by bending your knees (fig. 72 ②). Then drive your elbow into his midsection (fig. 72 ③). Continue the movement by reaching up with your right hand and grasping his right upper arm just above his elbow. With your left hand, grip his right wrist and throw him over your head (fig. 72 ④ and ⑤. He will hit the ground on his back presenting a good target to attack.

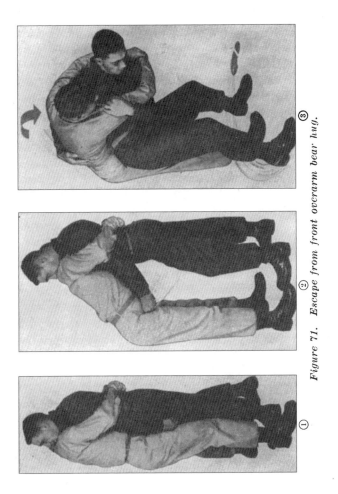

Figure 71. Escape from front overarm bear hug.

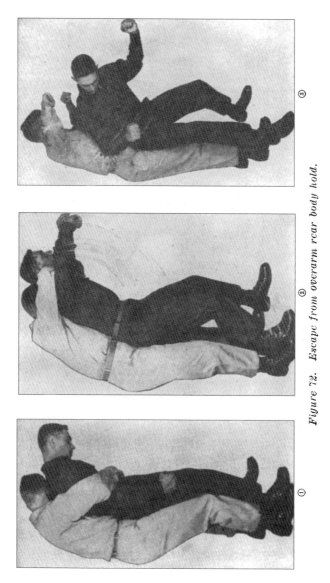

Figure 72. Escape from overarm rear body hold.

④

⑤

Figure 72.   Escape from overarm rear body hold—
Continued.

82. Escape From Front Underarm Bear Hug

Figure 73 ① shows that your opponent has locked his arms around your waist and is attempting to bend you over backward. Close the fingers of your left hand and place the thumb underneath the base of his nose. Put your right arm around his waist. By pressing with your left thumb and pulling his waist toward you, he either loosens his grip or is forced backward (fig. 73 ②).

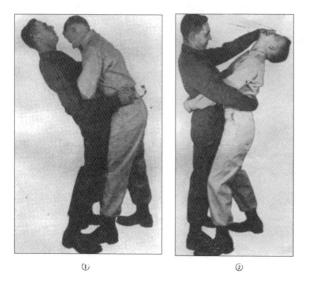

①   ②

Figure 73. Escape from front underarm bear hug.

83. Escape From Rear Underarm Body Hold

When your opponent grasps you as shown in figure 74 ①, reach down with your left hand and place it just above his left knee. Press down on this spot with most of your weight (fig. 74 ②). You now have a firm base on which to pivot. Lift both your feet

Figure 74. Escape from rear underarm body hold.

from the ground and switch your left leg behind your opponent's right leg. As soon as your feet are firmly planted on the ground, bring your left hand under your opponent's left knee and your right hand under his right knee (fig. 74 ③). Lift up and raise him off the ground (fig. 74 ④). If your opponent releases his hold, you can drive his head into the ground. If he keeps his hold, fall on him and force his head into the ground.

③

④

Figure 74.   Escape from rear underarm body hold—Con.

84. Second Escape From Rear Underarm Body Hold

Your opponent uses the same grasp around your waist as explained in paragraph 83, but this time he braces himself by placing one leg between your legs and putting his head behind your shoulder blade out of reach of your arms (fig. 75 ①). To break this hold, bend swiftly from the waist and grasp the ankle of the foot which he has between your legs (fig. 75 ②). Keep your hold on his ankle and straighten your body. This puts pressure on your opponent's knee, causing him to release his hold and drop on his back (fig. 75 ③). If your opponent keeps his hold, fall backward on top of him, driving your weight into his midsection.

85. Escape From Two-Hand Grip on One Wrist

If your opponent grasps your right wrist with both hands as shown in figure 76 ①, step forward with your right foot and bend both knees. Keep the trunk of your body upright and bring your right elbow close to your stomach. Reach across with your left hand and grasp your right fist (fig. 76 ②). By straightening your legs and pulling back with the power of your body and arms, you bring pressure on your opponent's thumbs, forcing him to release his hold. At the completion of the escape, you are in position to deliver a blow to your opponent's head or neck with the knife edge of your right hand (fig. 76 ③).

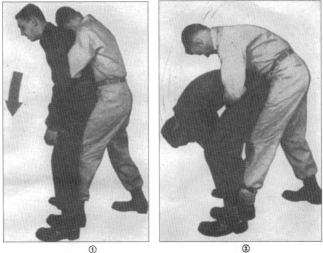

Figure 75. Second escape from rear underarm body hold.

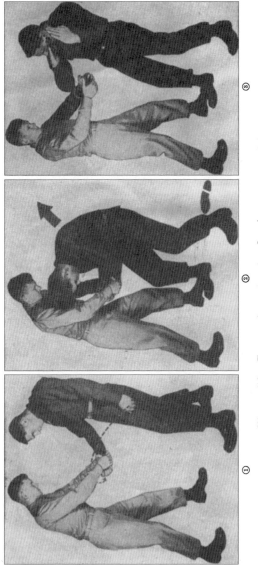

Figure 76. Escape from two-hand grip on one wrist.

86. Escape From Two-Hand Overhand Grip on Both Wrists

If your opponent grasps your wrists as shown in figure 77 ①, step forward with either foot and bend both knees. At the same time, bend your arms so the elbows are close to the lower abdomen (fig. 77 ②). Execute the escape by straightening your legs, pulling back with your body, and pushing your arms upward in one motion (fig. 77 ③). The faster you work this escape, the more effective it is.

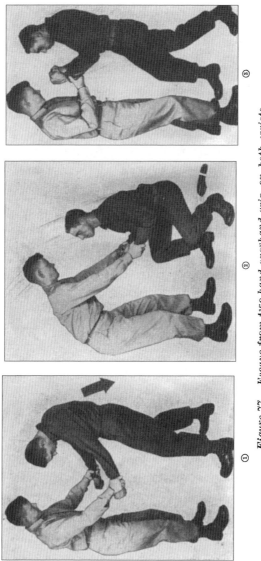

Figure 77. Escape from two-hand overhand grip on both wrists.

CHAPTER 8

DISARMING METHODS

Section I. BAYONET DISARMING

87. General

In training, you are taught bayonet disarming methods for both long thrust and short thrust attacks. In combat, however, any of the methods described in this section can be used to disarm an opponent whether he uses a long thrust or a short thrust. The reason the text differentiates between a short thrust and a long thrust is to enable you to gage the distance between the point of the bayonet and your body when practicing these techniques. When practicing the short thrust, you should be approximately arm's distance from the point of the bayonet. When practicing the long thrust, the distance is increased one foot. The unarmed man in training should wait until the armed man has committed himself before going into his disarming maneuver.

88. Counter Against Short Thrust

a. If your opponent attacks you with a short thrust, twist your body to the left but keep your feet in place. At the same time, slap your right forearm or wrist against the barrel of his rifle, deflecting the bayonet from your body (fig. 78 ①).

b. As soon as the bayonet has passed your body, grasp your opponent's left hand with your right hand. At the same time, take a long step with your left foot toward your opponent's right, reach under

①

Figure 78. Counter against short thrust.

the rifle with your left hand, and press your left
shoulder against the upper handguard. With your
left hand, grasp his right hand where it holds the
top of the small of the stock (fig. 78 ②).

c. Pull with your left hand and push with your
right hand. Keep your weight on your left foot and
kick your attacker with the calf of your right leg
behind the knee joint of his right leg (fig. 78 ③).
Your opponent will fall to the ground and loosen
his grip on the rifle.

89. Second Counter Against Short Thrust

a. As your opponent makes his thrust, use the heel
of your right hand to parry his bayonet to your
left and, at the same time, side-step to your right
oblique. You are now in a position facing the rifle
from the side with your groin area protected by your
right leg (fig. 79 ①).

②

③

Figure 78. Counter against short thrust—Continued.

Figure 79. Second counter against short thrust.

b. With your left hand, palm up, grasp the rifle on the upper handguard. At the same time, strike the inside of your opponent's left elbow sharply with the knife edge of your right hand (fig. 79 ②).

c. Keep a firm hold on the rifle. Step through with your left foot, moving quickly past your attacker on his left, and jerk the rifle up and backward in an arc over his shoulder (fig. 79 ③). If he keeps his hold on the rifle, kick him and yank the rifle loose. Whirl and attack him with the bayonet.

②

③

Figure 79. Second counter against short thrust—Continued.

90. Third Counter Against Short Thrust

a. As your opponent makes his thrust, use the heel of your left hand to parry the bayonet to your right and side-step to your left oblique. You are now in position facing the side of the rifle with your groin area protected by your left leg (fig. 80 ①).

①

Figure 80. Third counter against short thrust.

b. With your right hand, palm up, grasp the rifle anywhere on the upper handguard and with the left hand, palm down, grasp the receiver (fig. 80 ②).

c. Keep a firm hold on the rifle with both hands and step through with your right foot, moving quickly past your opponent. Jerk the rifle sharply up and backward in an arc over the attacker's shoulder and twist it out of his hands (fig. 80 ③). Whirl and smash him with the butt or attack him with the bayonet.

②

③

Figure 80. Third counter against short thrust—Continued.

91. Counter Against Long Thrust

a. As your opponent executes the long thrust, parry the bayonet to your left by slapping it with the heel of your right hand and side-step to the right oblique. You are now in a position facing the side of the rifle with your groin area protected by your right leg (fig. 81 ①). With your left hand, palm up, grasp your opponent's left hand and the rifle from underneath (fig. 81 ②). Twist your body to the left in front of your opponent and place your right leg in front of his body (fig. 81 ③).

①

Figure 81. Counter against long thrust.

b. With the right hand, palm down, grasp your opponent's left hand and rifle from above. Twist the rifle and pull your opponent across your right leg. At the same time, exert pressure with the right

②

③

Figure 81. Counter against long thrust—Continued.

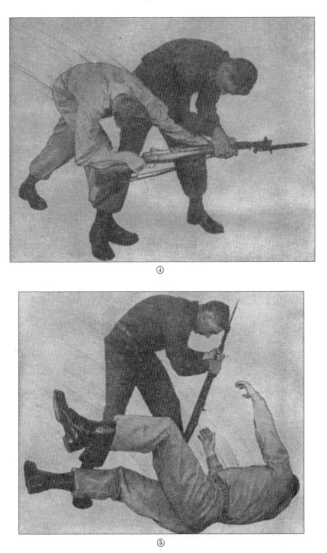

④

⑤

Figure 81. Counter against long thrust—Continued.

elbow against the outside of his left arm and elbow (fig. 81 ④). Sufficient pressure downward with your elbow, while twisting and pulling up on the rifle, can break your opponent's elbow.

c. Continue the twisting motion, pulling your opponent completely across your leg and throwing him to the ground (fig. 81 ⑤). Regrasp the rifle and follow through with an attack.

92. Second Counter Against Long Thrust

a. As your opponent executes the long thrust, parry his bayonet to your right with a sharp slapping movement with the heel of your left hand. As you parry with your left hand, move your body to the left oblique, stepping off to your left front with your left foot. You are now in position facing the rifle from the side with your groin area protected by your left leg (fig. 82 ①).

b. Strike the open palms of both hands down on the rifle near the muzzle, driving the point of the bayonet into the ground (fig. 82 ②). Do not follow the rifle all the way to the ground, but allow your opponent's momentum to imbed the bayonet into the ground.

c. Grasp the butt of the rifle with your left hand and with the right hand grasp your opponent anywhere on his back or head (fig. 82 ③). To completely disarm him, drive the stock of the rifle into your opponent's body and, at the same time, pull him with your right hand, spinning him to the ground (fig. 82 ④). You are now in position to recover the rifle and attack him.

①

②

Figure 82. Second counter against long thrust.

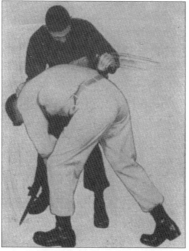

③

④

Figure 82. Second counter against long thrust—Continued.

93. Third Counter Against Long Thrust

a. This counter is essentially the same as the method described in paragraph 92 except that you parry left instead of right. This time as you parry, step to your right front with your right foot (fig. 83 ①).

①

Figure 83. Third counter against long thrust.

b. Execute the same movement as described in paragraph 92*b*. Use the open palms of both hands and drive the bayonet into the ground.

c. It may be difficult to reach across your opponent's body to grasp the butt of the rifle. Therefore, grasp his clothing with both hands and pull him forward, throwing him to the ground (fig. 83 ②).

②

Figure 83. Third counter against long thrust—Continued.

Section II. KNIFE DISARMING

94. Counter Against Downward Stroke

a. If your opponent attacks as shown in figure 84 ①, stop the blow by catching his wrist in the pocket formed by bending your fist forward at your right wrist. Step through with your right foot to protect your groin area. At the same time, strike him sharply in the crook of his right elbow with the thumb side of your forearm or wrist. This causes his arm to bend.

b. Bring your left hand behind his right forearm and underneath your right wrist, and grasp your right forearm. Bring your elbows close to your body (fig. 84 ②).

Figure 84. Counter against downward stroke.

c. Bend swiftly from the waist, putting pressure on your opponent's arm (fig. 84 ③). This causes him to fall backward and lose his weapon.

③

Figure 84. Counter against downward stroke—Continued.

95. Second Counter Against Downward Stroke

a. Stop the blow by catching your opponent's wrist in the pocket formed at your left wrist by bending your fist forward. Step through with your right foot to protect your groin area. Keep your left forearm horizontal to the ground. At the same time, bring your right hand underneath your opponent's knife arm and grasp your left fist (fig. 85 ①).

b. Bend swiftly forward from the waist and put pressure on your opponent's arm (fig. 85 ②). This causes him to fall backward and lose his weapon.

①

②

Figure 85. Second counter against downward stroke.

96. Counter Against Upward Stroke

a. Block an upward knife stroke by catching your opponent's wrist or forearm in the pocket formed at your left wrist by bending your fist forward. Keep your elbow low. At the same time, twist your body to the right (fig. 86 ①).

Figure 86. Counter against upward stroke.

b. As soon as you stop the blow, grasp your attacker's right hand with your right hand and place your thumb on the back of his hand. Reinforce this hold by grasping his wrist with your left hand and placing your left thumb on the back of his hand (fig. 86 ②).

c. Twist his wrist to your left and bend his hand toward his forearm, causing him to fall to the ground (fig. 86 ③).

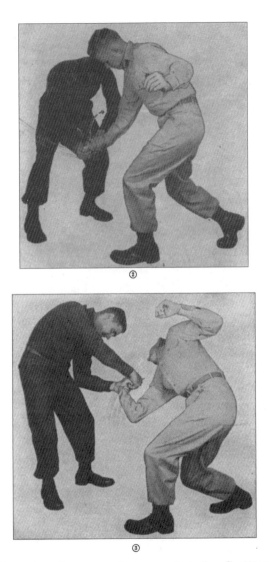

Figure 86. Counter against upward stroke—Continued.

97. Second Counter Against Upward Stroke

a. Another method of defending yourself against an upward stroke is to block your opponent's wrist or forearm in a "V" formed by your hands (fig. 87 ①). Keep your arms extended. Take a short crowhop to the rear as you block his thrust so your midsection is further from the point of the knife (fig. 78 ②).

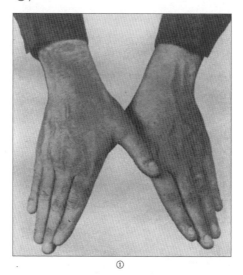

①

Figure 87. Second counter against upward stroke.

b. Grasp your opponent's wrist tightly with both hands and pivot to your left on the ball of your left foot. At the same time, raise your opponent's knife hand and step directly beneath his arm (fig. 87 ③). From this position, snap his arm forward and downward, bending at the waist and, at the same time, drive him to the ground with a whipping action (fig. 87 ④).

②

③

Figure 87. Second counter against upward stroke—
Continued.

④

Figure 87. Second counter against upward stroke—
Continued

98. Third Counter Against Upward Stroke

a. This phase is the same as that described in paragraph 97*a*.

b. Grasp your opponent's wrist tightly with both hands and pivot to your right. Raise his hand and bring his arm down over your left shoulder (fig. 88 ①).

c. Apply downward pressure on his arm. This will throw him to the ground or break his arm. This method can be varied by stepping completely under his arm and behind his back and forcing him to drop the knife by bending his arm (fig. 88 ②). You must pivot quickly to prevent him from spinning out of the hold.

①

②

Figure 88.   Third counter against upward stroke.

99. Counter Against Backhand Slash

a. To defend yourself against a backhand slash with a knife, bend your knees and lower your body without ducking your head. At the same time, raise your right arm and block your opponent's thrust with your forearm or wrist (fig. 89 ①).

b. As soon as you block the blow, grasp your opponent's knife hand with your left hand, your thumb in the center of the back of his hand. Apply pressure with your right wrist against his right wrist or forearm. Start to twist the knife hand to your left (fig. 89 ②), then reinforce your left-hand hold with a similar hold with your right hand. Both your thumbs are in the center of the back of his hand and your fingers are around his palm (fig. 89 ③). A twist to your left or pressure that bends your op

①

Figure 89. Counter against backhand slash.

②

③

Figure 89. Counter against backhand slash—Continued.

ponent's hand forward and under against his wrist causes him to lose his weapon and, in many cases, to suffer a broken or dislocated wrist.

100. Counter Against a Cautious Approach

When your opponent attacks as shown in figure 90 ①, his left foot is forward and his left hand is extended to ward off any of your possible blows. He holds his knife hand close to his right hip, ready to strike when an opening occurs. This is an extremely dangerous man. He is well-prepared and well-trained and your actions must be perfect. As soon as he comes within reach, spring from the ground, throwing your body at him feet first and twisting to your left. Hook your left instep around his forward ankle and kick his knee with your right foot (fig. 90 ②). Break the force of your fall with your hand or arm. This motion drops him on his

①

Figure 90. Counter against a cautious approach.

back. When both of you strike the ground, raise your right foot and kick his groin or midsection (fig. 90 ③).

Figure 90. Counter against a cautious approach—Continued

Section III. RIFLE DISARMING

101. Speed in Disarming

When disarming an opponent armed with a rifle or pistol, make each movement quickly and without hesitation. Although your opponent has the weapon, you are in a good position because you know what you are going to do whereas he has to react to your movement. *Although his reaction time is short, it is not as short as the time it takes you to act.*

102. Counter Against Rifle in Front

a. Figure 91 ① illustrates the holdup. At your opponent's order of "hands up," bring your hands to shoulder level. Then, in one motion, twist your body to your right and strike the muzzle of the rifle away

①

Figure 91. Counter against rifle in front.

②

③

Figure 91. Counter against rifle in front—Continued

from your body with your left forearm or wrist (fig. 91 ②).

b. As you strike the muzzle, step forward with your left foot and grasp the upper handguard with your right hand and the small of the stock with your left hand (fig. 91 ③).

c. Pull with your left hand and push with your right and step to your opponent's right with your own right foot. This knocks him off balance and, at the same time, enables you to strike him on the head with the muzzle of the rifle or to take the rifle from him by twisting it over his right shoulder (fig. 91 ④).

④

Figure 91. Counter against rifle in front—Continued

103. Counter Against Rifle in Back

a. When your opponent has his rifle in your back as shown in figure 92 ①, start to elevate your hands as ordered. When your hands reach shoulder height, twist from the hips to your right and bring your right elbow back, striking the muzzle of the rifle. This deflects the rifle away from your body. Do not as yet move your feet from their original position (fig. 92 ②).

①

Figure 92. Counter against rifle in back.

b. Turn to the right by pivoting on your right foot. Face your opponent and bring your right arm under the rifle and over your opponent's left wrist. Place your left hand on your opponent's right hand where it grasps the stock (fig. 92 ③). This prevents him from executing a butt stroke.

c. Pull with your left hand and push with your right shoulder and arm, forcing your opponent to the ground and making him release his grip on the rifle (fig. 92 ④).

②

③

Figure 92. Counter against rifle in back—Continued

133

Figure 92.   Counter against rifle in back—Continued

Section IV. PISTOL DISARMING

104. Counter Against Pistol in Front

a. Your opponent orders you to raise your hands. As you do so, keep your elbows as low as possible (fig. 93 ①). Twist your body to the right and strike your opponent's wrist with your left forearm (fig. 93 ②).

b. Grasp the bottom of the barrel with your right hand, making certain that your hand is not near the muzzle. At the same time, strike downward on your opponent's wrist with your left fist (fig. 93 ③). While applying pressure with your left fist, bend the pistol towards your opponent's body with your right hand, causing him to release his grip (fig.

①

Figure 93.   Counter against pistol in front.

93 ④). If he should retain his grip, his index finger will be broken. From this position, you can strike your opponent on his temple with the butt of the pistol.

105. Second Counter Against Pistol in Front

As you begin to raise your hands, bring them quickly forward and, at the same time, twist to your left away from the line of fire (fig. 94 ①). Bring your right hand under your opponent's wrist either with a grasping or a striking motion and, simultaneously, grasp the barrel of the pistol with your left hand (fig. 94 ②). Push up on the wrist with your right hand and down and out on the pistol with your left hand (fig. 94 ③). Your opponent will release his grip.

②

Figure 93. Counter against pistol in front—Continued

106. Counter Against Pistol in Back

a. This counter should be used only when you are certain that the pistol is in your opponent's right hand (fig. 95 ①). As you raise your hands, keep your elbows as close to the waist as possible. Twist your body to the right and, at the same time, bring your right elbow against your opponent's forearm

③

④

Figure 93. Counter against pistol in front—Continued

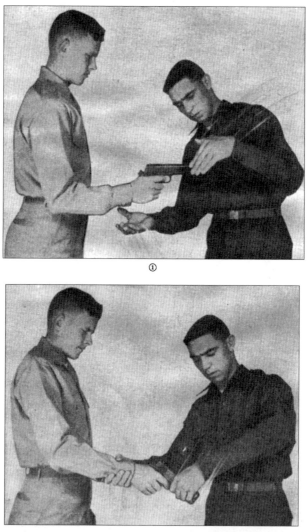

Figure 94. *Second counter against pistol in front.*

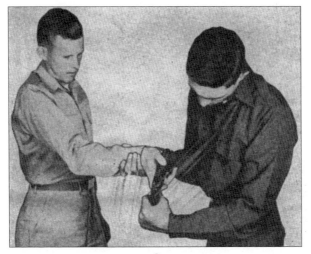

ⓐ

Figure 94. Second counter against pistol in front—Continued

①

Figure 95. Counter against pistol in back.

(fig. 95 ②). Keep your feet in place. Bring your right arm under your opponent's right forearm and place it on his elbow joint so that his forearm rests in the crook of your right elbow (fig. 95 ③).

b. Grasp your right hand with your left hand and bend swiftly from the waist (fig. 95 ④). By doing this, you force your opponent to the ground and cause him to drop his weapon.

107. Second Counter Against Pistol in Back

a. Use this counter when you are certain that your opponent is holding the pistol in his right hand. Keep your elbows as close to your waist as possible. Twist to the left, striking your opponent's wrist or forearm with your left elbow (fig. 96 ①). Bring your left arm behind your opponent's right elbow

②

Figure 95. Counter against pistol in back—Continued

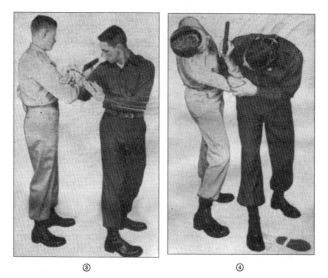

Figure 95. Counter against pistol in back—Continued

so that his forearm or wrist rests on your shoulder or neck (fig. 96 ②).

b. Grasp your left hand with your right hand and press your left forearm against your opponent's right elbow (fig. 96 ③). A swift twist to the front brings your opponent to the ground. With added pressure, you can break his arm. During the entire operation, the muzzle of the pistol is always pointed elsewhere.

108. Third Counter Against Pistol in Back

In this counter, it does not matter whether your opponent holds the pistol in his right or left hand because your actions are the same. The description given is for the pistol held in the left hand. Figure 97 ① illustrates the holdup.

①

②

Figure 96.   Second counter against pistol in back.

③

Figure 96. Second counter against pistol in back—Continued

a. Twist your body to the right striking your right elbow against your opponent's hand or wrist (fig. 97②).

b. Pivot to the right and place your left wrist against your opponent's left wrist, grasping the pistol barrel with your right hand, palm up. Apply pressure to his hand and trigger finger by pushing the barrel toward his upper arm. This releases his hold on the pistol and may break his index finger (fig. 97 ③). You now have the pistol in your right hand, opposite your left shoulder. By twisting forcefully to the right, you can strike your opponent on the chin or neck with the pistol butt (fig. 97 ④).

109. Counter Against Pistol in Back of Neck

a. This counter is practical only when you are certain that the pistol is held in your opponent's right

Figure 97. Third counter against pistol in back.

Figure 97. Third counter against pistol in back—Continued

hand. In raising your arms, bring your elbows shoulder high (fig. 98 ①). Twist your body to the left and bring your left arm under your attacker's right elbow (fig. 98 ②).

b. Reach across with your right hand and grasp your own left hand. Twist forward and put pressure on your opponent's elbow with your left fore-arm. You can either break his arm or force him to the ground, causing him to release his weapon (fig. 98 ③).

110. Second Counter Against Pistol in Back of Neck

Use this counter primarily for an attack with the pistol held in the right hand. The initial move, however, can be used for a right- or left-handed attack.

a. Hold the elbows shoulder high (fig. 99 ①).

①

②

Figure 98. Counter against pistol in back of neck.

Figure 98. Counter against pistol in back of neck—Continued

Twist your body to the right and, at the same time, bring your right upper arm over your opponent's right wrist (if the pistol should be in your opponent's left hand, bring it over his left wrist) (fig. 99 ②).

b. Pivot on your right foot and place your left foot close to your opponent's right foot. Hold your opponent's wrist close to your right side with your right upper arm. Cross your left arm under his right upper arm and grasp the left lapel of his shirt or jacket with your left hand (fig. 99 ③). Hold his right wrist close to your side and lift with your left upper arm, applying pressure to his elbow.

111. Third Counter Against Pistol in Back of Neck

Hold the elbows shoulder high (fig. 100 ①). The initial movement in this action is identical with that

①

②

Figure 99. Second counter against pistol in back of neck.

③

Figure 99. Second counter against pistol in back of neck—
Continued

shown in figure 99 ②. Twist your body to the right and strike your opponent's left wrist with your right arm (fig. 100 ②). Pivot on your right foot and place your left hand against your opponent's shoulder or upper arm. Bring your right forearm or wrist under your opponent's left elbow and lock it to your left forearm (fig. 100 ③). Apply pressure and cause your attacker to drop his weapon. Severe pressure can break his arm.

112. Helping a Friend

As you approach an unsuspecting opponent from his rear who is holding up a friend, carry your right hand low and your left hand shoulder high (fig. 101 ①). You must grasp your opponent with both

①

②

Figure 100.   Third counter against pistol in back of neck.

③

Figure 100. Third counter against pistol in back of neck—
Continued

hands at the same time. Your right hand, palm up, grasps the hand holding the pistol and lifts it, while your left hand pushes his right upper arm from behind (fig. 101 ②). Turn your body to the left and continue to apply pressure by pushing with your left hand and pulling his right hand backward (fig. 101 ③). This causes him to drop to the ground or suffer a dislocated shoulder.

113. Actions Against Opponent Who Is Holding You and Your Friend at Gun Point

The original position in this holdup is shown in figure 102 ①. The gunman is moving the muzzle of the pistol from your friend to you and back again. You are standing on your friend's left.

①

②

Figure 101. Attacking an unsuspecting gunman from his rear.

③

Figure 101. Attacking an unsuspecting gunman from his rear—Continued

a. As the weapon swings away from you, step forward with your left foot, place your left hand on the back of the attacker's gun hand, and push forcefully to his left (fig. 102 ②).

b. Take a step with your right foot and a quick long step with your left foot. Your movement brings you in front of your opponent, with your back to him. At the same time, twist his hand to his left, turning it so his palm is turned up and his right elbow comes in contact with your left armpit (fig. 102 ③). Bear down on his right elbow with your armpit and lift up on his hand, causing him to release his weapon or suffer a broken arm.

①

②

Figure 102. Actions when you and your friend are held up.

③

Figure 102. Actions when you and your friend are held up—
Continued

CHAPTER 9

PRISONER HANDLING

Section I. SEARCHING

114. General

If you capture a prisoner, you will not normally search him alone. Move him to the rear where he can be covered by another person while you search. In extreme circumstances, however, it may be necessary to make a thorough search unassisted. Two methods for a rifle search are presented in this section.

115. Rules for Searching

The rules you should follow when searching a prisoner are—

a. Indicate by inference, speech, and actions that you are completely confident and will fire if necessary.

b. Do not let your prisoner talk, look back, move his arms, or otherwise distract you.

c. Never attempt to search a prisoner until you have him in a position of extremely poor balance.

d. Don't move within arm's reach of your prisoner until you have him in a position of extremely poor balance.

e. If armed with a pistol, hold it at your hip in a ready position, and keep it on the side away from your prisoner. Change the pistol to your other hand when searching his other side.

f. When you have assistance, keep out of your partner's line of fire. One soldier conducts the search while the other remains far enough away to observe the prisoner at all times.

g. Don't relax your guard after you have completed your search.

116. Technique of Search

a. The "pat" or "feel" method of searching a prisoner will reveal most weapons and concealed objects. Search the prisoner's entire body, paying particular attention to his arms, armpits, back, groin area, and legs. Thoroughly pat the clothing folds around his waist, chest, and the top of his boots. Knives can be concealed on a string around the neck or taped to any area of the body. Take extreme caution when putting your hand in a prisoner's pocket or in the fold of a garment so he won't be able to clamp your arm and trip you over his leg.

b. After the initial search, a detailed search is made when the prisoner is moved to the rear. Force him to disrobe completely and examine his entire body from the soles of his feet to the top of his head.

117. Prone Method of Searching When Armed With a Rifle

Make the prisoner lie down on his stomach so that his arms are completely over his head and close together. His legs are also completely extended, feet close together. Place the muzzle of the rifle in the small of his back, keeping the gun upright. Grasp the rifle tightly around the small of the stock, index finger on the trigger (fig. 103). After searching his back, order him to turn over and repeat the process. Twist the muzzle into the prisoner's clothing

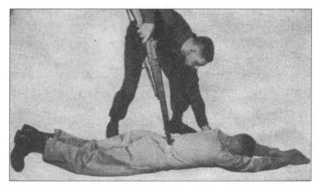

Figure 103. The prone method of searching makes it difficult for a prisoner to disarm his searcher.

to prevent it from slipping. You can also use this method when armed with a pistol. Keep the pistol at your hip while searching. You will have to use arm-and-hand signals for non-English speaking prisoners.

118. Kneeling Method of Searching When Armed With a Rifle

The prisoner interlocks his hands behind his head and kneels. He bends forward as far as possible until he is just able to maintain his balance. While searching his left side, hold the rifle in your right hand, muzzle jammed into the small of his back. Put your left leg between his legs and against his buttocks (fig. 104). In this position, you can quickly knock him flat by thrusting with the left knee. When searching his right side, hold the rifle in your left hand and put your right leg between his legs, knee against his buttocks. You can also use this method when armed with a pistol. Keep the pistol at your hip while searching.

Figure 104. The searcher keeps his left leg pressed into the prisoner's buttocks to knock the prisoner off balance, if the situation arises.

119. Wall Method of Searching When Armed With the Pistol (For Another Method of Wall Searching See FM 19–5)

Have the prisoner lean against a wall or a tree, one hand over the other and feet together and crossed as far to the rear as possible. This gives him extremely poor balance. To search his left side, place your left foot in front of his feet, keeping the pistol at your right hip (fig. 105). If the prisoner attempts to move, kick his feet out from under him. To search his right side, move to that side, switching the pistol to your left hand and placing your right foot in front of his feet.

Figure 105. The prisoner is kept off balance by leaning well forward with his legs and arms crossed.

120. Standing Method of Searching When Armed With a Pistol

Make the prisoner spread his legs far apart and place his hands on top of his head, fingers interlocked. While searching, keep as far from him as possible (fig. 106 ①). When you have to move close to him to search his front, place your foot against his heel, turning your body to the side to protect your groin (fig. 106 ②).

121. Searching More Than One Prisoner

a. A man armed with a pistol can search more than one prisoner at a time, using any of the methods presented in this section. While searching several prisoners, keep your eyes on all of them and do not look at your "patting" hand.

①

②

Figure 106. Search the prisoner well, but be alert at all times.

b. When using the wall method, keep the pistol in your right hand and search the left side of the first prisoner (fig. 107 ①). Step back and have him move to the far end of the line and resume the search position against the wall. Search the left side of the remaining men in the same manner. Then move to the other end of the line (fig. 107 ②). Hold the pistol in your left hand and search the right side of

①

②

Figure 107. When searching more than one prisoner, have each searched man move to the end of the line.

each prisoner, having each one move to the far end of the line as you finish.

c. To search more than one prisoner using the kneeling, prone, or standing methods, make them all assume the same position, in column about four or five feet apart (fig. 108). Search the rear man and then have him move to the front and resume the search position. Search the remaining prisoners from the rear, moving each to the front as you finish with him. When alone and armed with a rifle, the kneeling search is the best method to use.

Figure 108. The kneeling search is the best method if alone and when searching more than one prisoner.

Section II. SECURING

122. General

The most effective way of controlling a prisoner is to tie him. You can use pieces of clothing or equipment to tie him securely, such as shoe laces, leather or web belts, neckties, handerchiefs, or twisted strips of cloth. If a night patrol has the mission of capturing prisoners, it should always carry a piece of rope or flexible wire.

123. Belt Tie

Take the prisoner's belt and order him to lie on his stomach. Cross his arms behind his back and place the running end of the belt toward his feet, buckle toward his back. Hold the buckle on the wrist of his bottom arm and tightly wrap the running end around his wrist several times (fig. 109 ①). Place the running end of the belt parallel with his spine and outside the wrist of the upper arm. Now wrap the running end around the wrist of the upper arm several times (fig. 109 ②). Be sure to keep the prisoner's arms as close together as possible and to wrap the belt as tight as possible. Fasten the belt in the buckle (fig. 109 ③). Although this is an effective means of tieing, you should use it only when the prisoner is under close watch.

124. Shoe String Tie

Two 27-inch shoe strings or larger are needed for this tie. Have the prisoner remove his shoe or boot

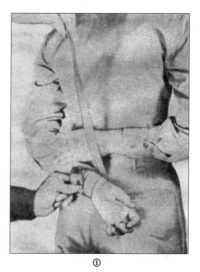

①

Figure 109. The belt tie.

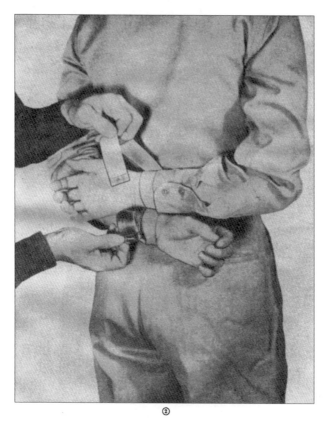

Figure 109. The belt tie—Continued

laces. You can make this tie with the prisoner's hands either in front of his body or behind his back, the latter being more effective. Place his hands back to back, wrists touching each other. Take one string and tightly wrap it completely around both wrists several times. Now wrap the end of the string around the cord between the insides of his wrist.

⑧

Figure 109. The belt tie—Continued

This will further tighten the outside loops. Tie the ends of the string together with any conventional knot (fig. 110 ①). Next tie the two little fingers together, using one end of the second string. Pass the remainder of the string over the loop around the wrists and tie his thumbs together. When you pass the second string around the wrists, be sure to pull it tight and keep it tight when tieing the thumbs (fig. 110 ②).

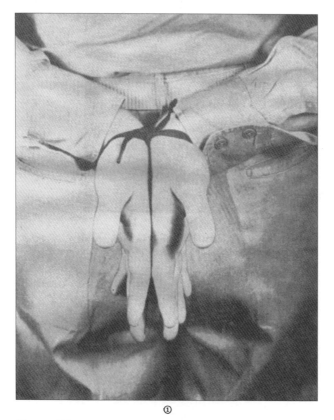

①

Figure 110. Many available objects can be used to tie a prisoner's hands, such as shoe strings.

125. Lead Tie

A piece of rope or two long boot laces are needed for this tie. Make the prisoner lie down. Tie his hands behind his back, using any conventional knot.

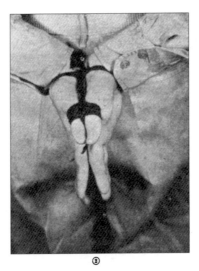

②

Figure 110. *Many available objects can be used to tie a prisoner's hands, such as shoe strings*—Continued

Figure 111. Lead tie.

Force his arms up behind his back in a strained position. Pass the rope around his neck and tie it around his wrists. The length of the loop around his neck should be short enough to force the man to keep his arms in a strained position to relieve pressure on his throat (fig. 111). The prisoner can be easily subdued by jerking on the rope as you walk behind him.

126. Hog Tie

Tie the prisoner in the lead tie (fig. 111). Cross his ankles and, after doubling his legs up behind him, tie them with the rope so that they remain in position. Any struggle to free himself will result in strangulation (fig. 112). When correctly applied, there is no escape from this tie.

Section III. GAGGING

127. Handkerchief

A gag prevents a prisoner from crying out. Force a handkerchief or a strip of cloth into the prisoner's

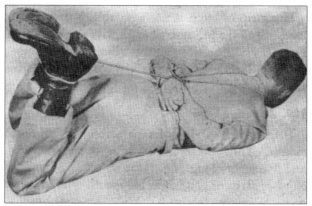

Figure 112. Hog tie.

mouth. A lump of turf will suffice if nothing else is available. Tie a handkerchief around the prisoner's mouth (fig. 113).

Figure 113.   Handkerchief gag.

128. Stick

If a strip of cloth is not available, a stick can be used. Stuff the prisoner's mouth with a piece of turf. Force the stick between his teeth like a bit in a horse's mouth. Tie the stick with a piece of shoe string around his neck (fig. 114).

129. Adhesive Tape

Place several strips of tape across the prisoner's mouth. The tape should be at least one inch wide and five inches long (fig. 115). Stuffing a handkerchief or strip of cloth into his mouth will make the gag much more effective.

Figure 114. Stick gag.

Figure 115. Adhesive tape gag.

CHAPTER 10

ADVICE TO INSTRUCTORS

Section I. THE INSTRUCTOR AND SAFETY PRECAUTIONS

130. The Instructor

The instructor should be in good physical condition and should be able to demonstrate all the maneuvers described in this manual. He must at all times display intense enthusiasm, vigor, and leadership qualities that will inspire the best efforts of the men being trained. He should train assistant instructors to closely supervise all practical work and to demonstrate new material.

131. General Precautions

The following safety precautions must be strictly followed during all training in unarmed combat—

a. Supervise all practical work closely and constantly. Never leave a class unsupervised.

b. Familiarize the students with each maneuver by complete explanation and demonstration before attempting any practical work.

c. During the learning stages, do not let the students get ahead of your instruction. This prevents accidents.

d. During the stages of learning and perfection of technique, the training partner offers no resistance. He should allow the maneuver to be freely executed.

e. Insure adequate space for all practical work. A space at least 8 feet wide by 8 feet long is needed for each pair of students.

f. Have the men empty their pockets before the practical work period and remove any jewelry, identification tags, etc.

132. Specific Precautions

The following precautions are applicable to the particular phases of training listed—

a. Vulnerable Points. Stress that only gentle blows are struck during the initial practical work. As students become more advanced, harder blows may be used. But students at all times must be cautioned against using excessive force, since serious injury could result.

b. Fall Positions, Throws, Holds.

    (1) Establish a signal that can be given by the individual student to stop the application of pressure when practicing holds. All students must know this signal, *particularly when practicing strangle holds.* This signal can, for example, be a clap of the hands or tapping your training partner.

    (2) Be sure all men are warmed up before doing any practical work.

    (3) Teach fall positions before conducting practical work in throws.

    (4) Caution the men to apply very light pressure until they become familiar with the hold.

c. Disarming.

    (1) While the men are learning bayonet disarming methods, keep scabbards on the bayonets and keep the bayonets firmly attached to the rifles.

    (2) Use tent pegs or bayonet scabbards to simulate knives in learning knife disarming.

    (3) Caution the men who are to be disarmed not to place their fingers in the trigger guards of their pistols or rifles while practicing disarming with these weapons.

Section II. CONDUCT OF TRAINING

133. Formations

a. Regulation physical training formations may be used for practice (FM 21–20). From the extended platoon formation, have the 1st and 3d ranks face the 2d and 4th ranks, so that each man will have a partner. Even numbered men do not uncover. It is recommended, when practicing throws, that twice the normal distance be taken between ranks.

b. For disarming methods, it is recommended that you use a formation of two concentric training circles or a formation that employs only two well-extended ranks. In forming the concentric training circle, pair the men off so that each will have a partner to work with.

134. Commands

a. Most of the maneuvers described in this manual can be divided into several steps or phases. To facilitate learning and to insure that the student learns each movement of an entire maneuver accurately, each maneuver is presented by phases.

b. For example, the right hip throw (par. 60) is a three phase maneuver. In the first phase, the student places his left foot in front of and slightly to the inside of his partner's left foot. At the same time, he strikes his partner on his right shoulder and grasps his clothing at this point. The command for this movement is PHASE ONE OF THE HIP THROW, MOVE. At the command MOVE, the student executes phase one and holds his position until given the subsequent command for the next movements. These commands are PHASE TWO (THREE), MOVE. When the students have become proficient in their movements, you can then work for speed. The phases of the maneuver are combined into a continuous movement by commanding HIP THROW, MOVE. At first, the maneuver is executed slowly. Students gain speed through constant practice.

c. Paragraph 92 explains the second counter against the long bayonet thrust. This disarming method is divided into three phases. For the first phase (par. 92*a*), the armed student is given the preparatory command LONG THRUST, and the unarmed student is given the preparatory command PARRY RIGHT. Since you want both students to halt their movements and remain in position for a subsequent command, you must command HOLD. The entire command for the first phase, therefore, is LONG THRUST, PARRY RIGHT, AND HOLD, MOVE. The next two phases of the movement are executed while the armed man is extended in the long thrust. The command for the second phase (par. 92*b*) is GROUND AND HOLD, MOVE. The command for the third phase (par. 92*c*) is DISARM AND HOLD, MOVE. The command for executing this disarming maneuver is DISARM FROM THE LONG THRUST, MOVE.

135. Exercise

a. Recommendations for Warm-up Exercises. Use combative exercises, grass drills, and tumbling exercises to warm up your men. The aggressive nature of these warm-up drills lend themselves to the spirit of hand-to-hand combat (FM 21–20).

b. Recommended Drill for Parry Exercise. It is important that your men be trained in the bayonet disarming parry movements before practicing the disarming methods. To do this, have the paired men assume their respective guard positions with the chin of the unarmed man 6 to 8 inches from the point of the bayonet. The armed man stands fast in the guard position during the entire exercise. The unarmed man parries the bayonet first to the right and then to the left, on command. As he parries to the right, he side-steps to his left oblique with his left foot, brings his right foot slightly to the rear of his left, and faces the side of the rifle. When he parries left, he side-steps with his right foot to his right oblique.

Section III. THE TRAINING AREA

136. Training Pit

a. The most suitable area for teaching fall positions, throws, and counters is the sawdust pit. You can get sawdust easily at most stations. Figure 116

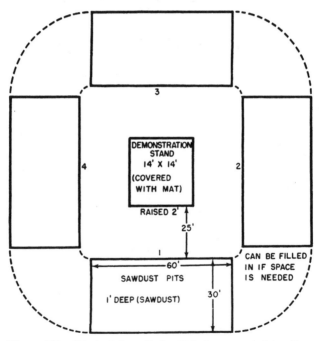

Figure 116. The training pit should be large enough to allow maneuvering by the students.

shows such an area. Each pit will accommodate twelve pairs of men. If you need additional room, fill in the area to the sides of each pit, as shown.

b. To construct a pit, either dig out 12 inches of ground or build a retaining wall of dirt about 12 inches high around the sawdust pit.

c. Place a raised platform, about 14 feet square, in the center of the pit area for demonstrations. Cover it with a removable mat. This platform is big enough to hold both the demonstrators and principal instructor during the demonstration.

d. The men gather around the platform to watch the demonstrations.

137. Other Areas

a. Any large grassy or sandy area is suitable for work in disarming methods and throws. Students should be taught fall positions before learning throws on a hard grassy area.

b. Classes in unarmed combat can also be conducted indoors. In this case, the floor and walls should be matted. The difficulty here is that only a small group of men can participate at one time, since an area eight feet square should be allotted for each two men.

138. Variation in Minimum Training Program

Periods eight and nine of the subject schedule include work in throws, falls, and escapes from basic holds. This time, however, may be spent in reviewing previous lessons, if the instructor thinks the students need review. The throws, falls, and escapes can be included in the physical training program.

Section IV. MINIMUM TRAINING PROGRAM*

139. Subject Breakdown

Period	Subject	Type	Area	General subjects department	
				Student equipment	References
1	INTRODUCTION, basic fundamentals attacking vulnerable points. STRANGLE HOLDS.	C, D, PE	Field	One tent peg per two students.	FM 21–150, chs. 1, 2, 3, 7.
2	Review Vulnerable Points and Strangles, silencing Sentries, HOLDS.	C, D, PE	--do--	One tent rope and one steel helmet per two students.	FM 21–150, chs. 5 & 7.
3	Review PE Strangle Holds and Silencing Sentries. KNIFE ATTACK.	C, D, PE	--do--	One tent rope, 1 tent peg or bayonet scabbard, and 1 steel helmet per two students.	FM 21–150, ch. 4.
4	Review Knife Attack, Bayonet Disarming.	C, D, PE	--do--	One tent peg and 1 rifle w/bayonet and scabbard per two students.	FM 21–150, section I, ch. 8.

5	Review Bayonet Disarming, Knife Disarming.	C, D, PE	Field----	One tent peg and 1 rifle with bayonet and scabbard per two students.	FM 21–150, section II, ch. 8.
6	Review Knife Disarming, Pistol Disarming.	C, D, PE	---do----	One tent peg and 1 wooden pistol per 2 students.	FM 21–150, section IV, ch. 8.
7	Review Pistol Disarming, Rifle Disarming.	C, D, PE	---do----	One wooden pistol and 1 rifle per 2 students.	FM 21–150 section III, ch. 8.
8	Side Fall Position, Hip Throw, Reverse Hip Throw.	C, D, PE	Pits----	None----------	FM 21–150, ch. 6.
9	Over Shoulder Fall Position, Over Shoulder Throw, Escapes from Holds.	C, D, PE	---do----	None----------	FM 21–150, ch. 6 & 7.
10	Prisoner Handling----------	C, D, PE	Field----	One belt, 1 pair shoe strings, 1 rifle, 1 wooden pistol, one 15-foot rope per 2 students.	FM 21–150, ch. 9.

*This program is used to give men a brief orientation in unarmed combat. Much more time must be spent to take men proficient. If more time is available, suggest the use of applicable portions of TF 19–1634, Personal Encounters.

APPENDIX

REFERENCES

1. Publications Indexes

Special Regulations in the 310–20-series; SR 110–1–1; and FM 21–8 should be consulted frequently for latest changes or revisions of references given in this manual and for new publications relating to subject matter covered in this manual.

2. Other Publications

SR 320–5–1	Dictionary of United States Army Terms.
SR 320–50–1	Authorized Abbreviations.
FM 19–5	Military Police.
FM 21–5	Military Training.
FM 21–20	Physical Training.

INDEX

[AG 353 (11 Mar 54)

By order of the Secretary of the Army:

M. B. RIDGWAY,
General, United States Army,
Official: *Chief of Staff.*

JOHN A. KLEIN,
Major General, United States Army,
The Adjutant General.

Distribution:

Active Army:

Tech Svc (1); Tech Svc Bd (1); AFF (15); AA Comd (2); OS Maj Comd (5); Base Comd (5); MDW (5); Log Comd (5); A (10); CHQ (5); Div (5); Brig (5); Regt 5, 7, 17 (5); Bn 5, 6, 7, 17, 19, 44 (5); CO 5, 7, 17, 19, 57 (10), 6 (2); FT (2); USMA (50); Sch (5) except 7; PMS & T (1); Tng Div (10); Mil Dist (3); Mil Mis (1); Arma (1).

NG: Same as Active Army except two copies to each unit.

USAR: Same as Active Army except two copies to each unit.

For explanation of distribution formula, see SR 310–90–1.